SINGLE VARIABLE
CalcLabs
WITH MAPLE®

for Stewart's

FOURTH EDITION

CALCULUS
SINGLE VARIABLE CALCULUS
CALCULUS: EARLY TRANSCENDENTALS
SINGLE VARIABLE CALCULUS: EARLY
  TRANSCENDENTALS

Al Boggess
David Barrow
Maury Rahe
Jeff Morgan
Philip Yasskin
Michael Stecher
Art Belmonte
Kirby Smith

Texas A & M University

BROOKS/COLE PUBLISHING COMPANY

I⊤P® An International Thomson Publishing Company

Pacific Grove • Albany • Belmont • Bonn • Boston • Cincinnati • Detroit • Johannesburg • London
Madrid • Melbourne • Mexico City • New York • Paris • Singapore • Tokyo • Toronto • Washington

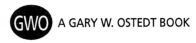 **A GARY W. OSTEDT BOOK**

Assistant Editor: *Carol Ann Benedict*

Marketing Manager: *Caroline Croley*

Marketing Assistant: *Debra Johnston*

Production Coordinator: *Dorothy Bell*

Cover Illustration: *dan clegg*

Printing and Binding: *West Publishing*

For more information, contact:

BROOKS/COLE PUBLISHING COMPANY
511 Forest Lodge Road
Pacific Grove, CA 93950
USA

International Thomson Editores
Seneca 53
Col. Polanco
11560 México, D. F., México

International Thomson Publishing Europe
Berkshire House 168-173
High Holborn
London WC1V 7AA
England

International Thomson Publishing GmbH
Königswinterer Strasse 418
53227 Bonn
Germany

Thomas Nelson Australia
102 Dodds Street
South Melbourne, 3205
Victoria, Australia

International Thomson Publishing Asia
60 Albert Street
#15-01 Albert Complex
Singapore 189969

Nelson Canada
1120 Birchmount Road
Scarborough, Ontario
Canada M1K 5G4

International Thomson Publishing Japan
Palaceside Building, 5F
1-1-1 Hitotsubashi
Chiyoda-ku, Tokyo 100-0003
Japan

Printed in the United States of America

10 9 8 7 6 5 4 3

ISBN 0-534-36433-0

Contents

Introduction

Maple is a powerful software tool for mathematical computations and visualization. Maple is radically changing the way scientists and engineers do mathematics in much the same way that calculators changed the computational landscape in the 1970's. The goal of this manual is to introduce this software package to students who are simultaneously taking first year calculus and using the fourth edition of James Stewart's **Calculus.** Our focus is on the use of Maple as a tool to solve problems that are difficult to solve with hand computation. In addition, we hope that by using Maple to solve problems, students will also improve their understanding of the concepts of calculus.

This manual is written for a calculus course that involves at least one computer laboratory contact hour per week (with some additional open laboratory hours available outside of class). Many large universities cannot offer more than one contact hour per week in a computer laboratory due to budget constraints and the ever-increasing enrollments in calculus. For this reason, the goals of this manual are rather modest. The manual uses relatively few Maple commands in order to keep the amount of syntax to a minimum. Most of the examples and exercises involve little formal programming with Maple. Indeed, one of the beautiful qualities of Maple is that much can be done with few commands. As another disclaimer, this manual will not magically elucidate all the concepts of calculus (although we hope it will help). Students of calculus must still rely on good lectures by instructors (unless the class size is small enough to allow a more radical approach) and study efforts that involve lots of homework problems (some with Maple and some without).

The first part of this manual (chapters 1 - 9) reads like a standard text, introducing the use of Maple with examples. Chapters 1 and 2 serve as introductory chapters which should acquaint students with some essential Maple commands. Chapters 3 through 8 are closely related to the same-numbered chapters in Stewart's text, and chapter 9 is associated with Stewart's appendix on Polar Plots. Sections 5.5 and 6.3 are noted exceptions. References are given to Stewart's text at the beginning of each section, and exercises are given at the end of each chapter.

The second portion of this manual contains Maple programming information, trouble shooting tips and 20 projects. Chapter 10 contains material on Maple programming as well as descriptions of some of the more sophisticated Maple commands. Chapter 11 contains a number of helpful trouble shooting tips. Chapter 12 contains twenty student projects of varying levels of difficulty. Most of the projects involve some area of application where Maple is used in the problem-solving process to simplify computations and to help with graphics.

The exercises and examples in this manual vary considerably in length and difficulty. Some of the exercises and examples are routine and designed to illustrate Maple syntax. Others are more involved and are designed to solve problems or illustrate ideas that would be difficult using only hand computation. Many of the more complicated examples and exercises are designed to embellish classical problems with more real-life considerations. The following example (presented in section 4.6) illustrates this aspect. Consider the problem of minimizing the surface area of a cylindrical can of fixed volume. Most calculus books in print over the past several generations contain this problem as an example or an exercise. After this problem has been translated into mathematics, its solution is straight forward and does not require the use of a computer. However, the computer allows the student to consider a more real-life version of this problem that seeks to minimize the cost of constructing this can where the cost of the seam used to attach the top, bottom and sides is considered along with the cost of the material. Minimizing this cost function requires the use of the computer for computations and graphics. Students should be expected to do the classical version of this problem by hand, and then they can be assigned the more real-life version of this problem in the computer laboratory.

As a word of warning to students, Maple alone will not solve calculus problems. "Thinking" is required for problem set-up. Maple can only be used to help with the computations and graphics that are necessary to obtain final answers. With this in mind, students should read the text of each chapter and set-up any assigned problems before coming to their computer lab in order to make the best use of time spent in front of the computer.

All of the Maple syntax and examples presented in this manual use Maple V Release 5. Most, but not all, of the syntax presented in this manual is valid for earlier versions of Maple. One exception involves the syntax of the **dfieldplot** command in chapter 7. In addition, the Maple command for the previous output changed from " to % in Release 5.

This manual assumes that the reader is using a windows platform (e.g. *X/Motif, Microsoft Windows*, or *Macintosh*), however references to platform-specific features of Maple (e.g. window menus) have been avoided in order not to tie this manual to any specific computer platform

Historical Information

This manual was originally written in the spring of 1995 with Maple V Release 3 and then exported as LaTeX files in order to ensure that no errors (hopefully) were made in transcribing Maple output into the text. It was revised in the spring of 1997 to work more closely with James Stewart's **Calculus: Concepts and Contexts, Single Variable**, and revised again in the spring of 1999 to conform to the syntax of Maple V Release 5 and the fourth edition of Stewart's **Calculus**.

All author royalties on copies of this manual sold to Texas A&M University students, and a portion of the author royalties on all other sales of this manual, have been donated to an undergraduate scholarship fund for Texas A&M University students.

Please send all comments and corrections to
Jeff Morgan, Department of Mathematics, Texas A&M University,
jmorgan@math.tamu.edu.

Syllabi

As mentioned in the Introduction, this manual is designed for students in a computer laboratory environment who are simultaneously taking a calculus course which is using James Stewart's **Calculus**. The following syllabus assumes that students have access to one hour per week of organized computer laboratory instruction together with two or three hours per week of outside access to a computer during the course of a fourteen week semester. The organized hour of computer laboratory can be conducted in a self-paced mode with students using the computer to try out the examples in the text. The laboratory instructor can circulate to answer questions or provide organized guidance on troublesome syntax issues (such as the distinction between expressions and functions). Students can then be assigned exercises or projects that can be started in the laboratory and completed outside of class.

Calculus I Syllabus

Chapters 1 through 4, and portions of chapter 5.

The instructor may choose to incorporate one or two of the projects from chapter 12. Any of the first eight projects (before the *Calculus I Review*) can be done with standard Calculus I tools. However, we advise caution here. Do not get too ambitious during the first semester since students need some time to adjust to Maple syntax and to Calculus (and to all the other life-changing attributes of college life). The assignment of one or two of the first four projects (which are the easier ones) is probably sufficient for those who plan to incorporate projects into Calculus I. The more ambitious projects can be postponed to Calculus II.

Calculus II Syllabus

The remainder of chapter 5 and chapters 6 through 8.

At this point, the students should have a small level of Maple sophistication. Consequently, the instructor might want to assign some reading from chapter 10. The instructor should also assign some of the projects in chapter 12. The projects are ordered according to the level of calculus required. The level of difficulty generally increases in the later projects (although this is not strictly true). The beginning of each project contains a brief statement on the concepts from calculus that are required. Some hints and guides as to how Maple can be utilized are sometimes given (especially for the earlier projects).

Basic Maple Commands

All Maple commands must be terminated with a semicolon (if output is desired) or a colon (to suppress output). Help on the syntax of any Maple command can be obtained by typing **?command**. For example, to get help with the **solve** command, type **?solve**.

A portion of the Maple commands used in this manual are listed below:

a:=1.53;   Assigns the value 1.53 to the variable a.

a:=Pi*r^2;   Assigns the name a to the given formula.

a:='a';   Unassigns any value previously given to a.

f:=x^2+5;   Defines the *expression* $f = x^2 + 5$.

f:=x->x^2+5;   Defines the *function* $f(x) = x^2 + 5$.

evalf(expr);   Evaluates **expr** as a floating point number.

expand(x^2*(2*x+1)^3);   Distributes multiplication over addition.

expand(sin(x+y));   Uses the formula for the sine of the sum of two angles to get $\sin(x)\cos(y) + \sin(y)\cos(x)$.

expand(exp(2*x));   Expands e^{2x} to the product $e^x e^x$.

simplify(expr);   Algebraically simplifies **expr**.

factor(expr);   Factors a polynomial **expr**.

solve(equation,x);   Tries to give an *exact* solution listing all x's which solve an equation. If the equation is a polynomial of degree three or more, the solution may be expressed as **RootsOf**. Maple may not know how to find *exact* solutions in some cases.

%   Refers to the output of the previous command.

sol:=solve({eqn1,eqn2},{x,y});   Gives *exact* solutions x and y to the system of simultaneous equations labeled eqn1 and eqn2. If there is more than one solution, they are **sol[1]** and **sol[2].**

assign(sol);   Assigns the values found for x and y above to those variables.

rhs(eqn); and lhs(eqn);   Read off the right and left sides of an equation, **eqn.**

fsolve(eqn,x=a..b);   For x in the interval $a \le x \le b$, seeks *floating point decimal* solutions to an equation, **eqn.** The **fsolve** command usually finds a solution if there is one, but it frequently finds only some of the solutions. Restricting the interval on which it searches will help it to find the other roots.

Pi;   The *exact* constant π.

exp(x);   The exponential function, e^x.

exp(1);   **The exact value of the constant** e.

plot(expr,x);   Plots **expr** over the default interval $-10 \leq x \leq 10$. The scale of the y-axis is adjusted to fit the y values that appear in the plot. The scales for the x-axis and the y-axis need not match.

plot(expr,x=a..b);   Plots **expr** over the interval $a \leq x \leq b$.

plot(expr,x=a..b,y=c..d);   Plots **expr** over the interval $a \leq x \leq b$, but restricts the displayed values of y to the range $c \leq y \leq d$.

plot(expr,x,scaling=constrained);   Plots **expr** over the default interval $-10 \leq x \leq 10$ using the maximum range of y's to determine the y-scale, but adjusting the x scale to correspond to the scale given for y so that there is less distortion.

plot({expr1,expr2},x);   Plots the graphs of **expr1** and **expr2** on the same coordinate axes.

plot(f,2..3);   Plots the *function* f over the interval $2 \leq x \leq 3$. Note that when a *function* is plotted, as opposed to an *expression*, no "x" is included in the plot command.

with(plots):   Loads the **plots** package that is necessary for the **implicitplot** command.

implicitplot(x^2+y^2=1,x=-1..1,y=-1..1);   Plots the *equation* $x^2 + y^2 = 1$.

plot([f(t),g(t),t=a..b]);   Plots the graph of the parametric equations $x = f(t)$ and $y = g(t)$ for $a \leq t \leq b$.

subs(x=a,expr);   Substitutes a for x at each occurrence of x in **expr**. The value of a could be a number or an algebraic expression like $2y + 3$.

Caution: The command **subs(x=a,expr);** *only* substitutes $x = a$ into the expression *expr* and does *not* change any previously assigned value of x.

Limit(expr,x=a);   Displays (but does not evaluate) $\lim_{x \to a} expr$.

value(%);   Computes the value of the previous output.

limit(expr,x=a);   Evaluates the limit of **expr** as $x \to a$.

limit(expr,x= infinity);   Evaluates the limit of **expr** as $x \to \infty$.

diff(expr,x);   Differentiates the *expression* **expr** with respect to x. The x is required although there may be no other variables in **expr**.

D(f);   Returns the derivative of the function f as a function.

Int(expr,x,x=a..b);   Displays (but does not evaluate) $\int_a^b expr\,dx$.

value(%);   Computes the value of the previous output.

int(expr,x);   Computes the indefinite integral of the *expression* **expr** with respect to x. The answer is another *expression*. Note that the x is required, even though there may be no other variables in **expr**.

int(expr,x=a..b);   Computes the definite integral of **expr** with respect to x over the interval $a \leq x \leq b$.

Sum(expr,i=m..n);   Displays (but does not evaluate) $\sum_{i=m}^{n} expr$.

value(%):   Computes the value of the previous output.

sum(expr,i=m..n);   Sums the *expression* **expr** as i goes from m to n.

p:=taylor(f,x=a,n);   Assigns to p the Taylor expansion of the expression f about the point $x = a$ with remainder of order n.

p:=convert(p,polynom); Converts p to a polynomial (this is useful to remove the remainder term appearing in a Taylor expansion).

with(student): Loads the student package that is necessary for certain commands such as **intparts** and **changevar**.

changevar(3*x=tan(theta),A,theta); Performs the change of variables $3x = \tan(\theta)$ in the integral labeled A (this command requires the student package to be loaded first—type **with(student):**).

intparts(A,uexpr); Performs one integration by parts on the integral labeled A with **uexpr** used as the u variable (this command requires the student package to be loaded first—type **with(student):**).

dsolve(eqn, y(t)); solves the differential equation labeled **eqn** for the unknown function $y(t)$.

dsolve({eqn, init}, y(t)); solves the differential equation labeled **eqn** with the initial condition labeled **init** for the unknown function $y(t)$.

Chapter 1

Getting Started

Background Information: Precalculus.

This chapter introduces some of the basic Maple commands associated with assigning variables and creating plots.

1.1   Maple as a Calculator

A Maple input line is indicated by an input prompt > at the left hand margin. A Maple command is entered by typing it on an input line with a semicolon (;) at the end and then pressing the ⟨ENTER⟩ or ⟨RETURN⟩ key. Try entering

```
> 2 + 5;
                                7
```

(Don't type the > prompt since this is provided by the computer.) Maple's output, 7, is displayed between the lines in the center of the page as it appears on the computer screen.

If a command is entered and Maple does nothing after ⟨ENTER⟩ is pressed, then a semicolon is probably missing. Type a semicolon and press ⟨ENTER⟩. If a command or an expression is entered incorrectly then click the mouse on the line to edit it and then re-execute it by pressing ⟨ENTER⟩ again.

Maple can do arithmetic on formulas entered on an input line. The standard arithmetic operations are

+	addition
-	subtraction
*	multiplication
/	division
^	exponentiation

1

The standard order of operations is exponentiation before multiplication and division and then addition and subtraction. To be safe, use parentheses to be sure that the operations are performed in the desired order. For example, **(3+4)/7;** is not the same as **3+4/7;**. Be sure to use round parentheses rather than square brackets [ ] or curly braces { }, which have other meanings in Maple. **Caution:** A typical mistake is typing

```
> (2+4)(3-1);
```

instead of

```
> (2+4)*(3-1);
```

Maple will not multiply without the ***** sign. Rather it will give very peculiar results without warning.

Maple knows a large number of standard mathematical functions including:

the square root function	**sqrt**
the absolute value function	**abs**
the natural exponential	**exp**
the natural logarithm	**ln**
the trig functions	**sin, cos, tan, sec, csc, cot**
the inverse trig functions	**arcsin, arccos, arctan,**
	arcsec, arccsc, arccot

It is important to understand the distinction between numbers that Maple knows exactly, such as 2, 1/3, $\sqrt{2}$, and π, and *floating-point decimal numbers* such as 2.0, .33333, 1.414, and 3.14. The number 1/3 is an expression that represents the exact value of one-third, whereas .33333 is a floating-point decimal approximation of 1/3.

If numbers are entered as integers, then Maple normally returns exact answers. For example, enter

```
> (1+3)/6;
```
$$\frac{2}{3}$$

Maple returns 2/3 rather than its decimal approximation .6666666667. To get the decimal approximation, use the **evalf** command. For example, try

```
> evalf(22/79+34/23);
```
$$1.756741882$$

Notice that parentheses are needed around the expression. Here, **evalf** means to evaluate the expression as a floating-point decimal number. Alternatively, one of the numbers can be entered as a decimal. For example, type

```
> 22./79 + 34/23;
```

$$1.756741883$$

and Maple returns the answer in decimal form.

1.2 Assigning Variables

Maple answers are often used again in subsequent calculations and therefore Maple provides a way to store and recall earlier results. One way to refer to an earlier result is to use the ditto mark **%**, which refers to the immediately preceding result. For example, the calculation of $2^6 + 1$ can be done in two steps to show intermediate answers. The result of the first operation is given to the second part as **%**.

```
> 2^6; %+1;
```

$$64$$

$$65$$

Note that the input line contains two Maple commands (since there are two semicolons) and therefore there are two Maple outputs.

Names or labels can also be used to store and refer to results. For example, the number 22./79+34/23 can be assigned to the variable a by typing

```
> a:=22./79+34/23;
```

$$a := 1.756741883$$

Maple commands of this form are called assignment statements and the := sign indicates that the quantity on the right is to be assigned to the variable name on the left. Now, the number 1.756741883 can be recalled by typing **a;**.

Examples. To compute $(1.756741883)^2$, type

```
> a^2;
```

$$3.086142043$$

To compute $\dfrac{1}{1.756741883}$, enter

```
> 1/a;
```
$$.5692355887$$

Compute $\sqrt{1.756741883}$

```
> sqrt(a);
```
$$1.325421398$$

To more easily distinguish between various labels, use descriptive names. For example, the retail price and wholesale cost of an item can be stored by using the assignment statements

```
> Price:=4.95;
```
$$Price := 4.95$$

```
> Cost:=2.80;
```
$$Cost := 2.80$$

The value of the profit is then given by

```
> Profit:=Price-Cost;
```
$$Profit := 2.15$$

Labels are case sensitive: the label *Profit* is different than the label *profit*.

A variable keeps its value until it is assigned a new value or until it is cleared (or unassigned). The value from the assigned variable *Cost* can be cleared by entering

```
> Cost:='Cost';
```
$$Cost := Cost$$

Now the variable *Cost* has no value assigned to it.

As another example, enter an expression that describes the area of a circle of radius r

```
> Area:=Pi*r^2;
```
$$Area := \pi\, r^2$$

Note that π is entered with an upper case P. With a lower case p, Maple will show the Greek letter π but won't recognize its mathematical meaning. To evaluate this area when $r = 5$, enter **r:=5;**. The value $r = 5$ will automatically be substituted into *Area*.

> **r:=5; Area; evalf(%);**

$$r := 5$$

$$25\,\pi$$

$$78.53981635$$

1.3 Algebra Commands

We have seen how to manipulate numbers and assign them to variables (or labels). Maple can also manipulate algebraic expressions involving labels or variables.

For example, to multiply out the expression $(3x - 2)^2(x^3 + 2x)$, type

> **expand((3*x-2)^2*(x^3+2*x));**
$$9\,x^5 + 22\,x^3 - 12\,x^4 - 24\,x^2 + 8\,x$$

It is easy to make a typographical error when typing a complicated expression, such as $(3x - 2)^2(x^3 + 2x)$. To prevent such errors from affecting a Maple command (such as **expand**), first type the expression without the command (followed by the ⟨RETURN⟩ key) to see if it is entered correctly. Then insert and execute the Maple command. For example, the above expansion can be performed as follows: first type

> **(3*x-2)^2*(x^3+2*x);**
$$(3\,x - 2)^2\,(x^3 + 2\,x)$$

and examine Maple's output to make sure that the expression is entered correctly. Then click the mouse back at the end of the previous line and type **expand(%);**.

> **(3*x-2)^2*(x^3+2*x); expand(%);**
$$(3\,x - 2)^2\,(x^3 + 2\,x)$$

$$9\,x^5 + 22\,x^3 - 12\,x^4 - 24\,x^2 + 8\,x$$

As mentioned earlier, the percent **%** refers to the output preceding the quote—in this case, the expression $(3x - 2)^2(x^3 + 2x)$.

To factor the polynomial $x^6 - 1$, type

```
> x^6-1; factor(%);
```

$$x^6 - 1$$

$$(x - 1)(x + 1)(x^2 + x + 1)(x^2 - x + 1)$$

Another useful command is **simplify**. For example, to simplify the expression

$$\frac{x^2 - x}{x^3 - x} - \frac{x^2 - 1}{x^2 + x}$$

enter

```
> (x^2-x)/(x^3-x)-(x^2-1)/(x^2+x); simplify(%);
```

$$\frac{x^2 - x}{x^3 - x} - \frac{x^2 - 1}{x^2 + x}$$

$$-\frac{x^2 - x - 1}{x(x + 1)}$$

The following command will also simplify the expression.

```
> simplify((x^2-x)/(x^3-x)-(x^2-1)/(x^2+x));
```

$$-\frac{x^2 - x - 1}{x(x + 1)}$$

However, with this syntax, it is harder to keep track of the parentheses in such a long expression. In addition, this command does not display the original expression and therefore it cannot be checked for typing errors.

Note: Maple has an on-line Help facility that is invoked by typing **?**. Help with a specific command can be obtained by typing a **?** followed by the command. For example, to get help with the **factor** command, type **?factor**. Also explore the Help Browser and the new Keyword Search facility by clicking on Maple's **Help** menu.

1.4   Plots

The **plot** command is best introduced with an example. To plot the graph of

$$y = \frac{2x^2 - 4}{x + 1}$$

over the interval $-6 \le x \le 6$, type

> **plot((2*x^2-4)/(x+1),x=-6..6);**

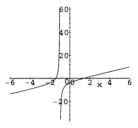

The scale on the y-axis is much different from the scale on the x-axis because of the very large function values when x is close to -1 (where the function becomes undefined). To get a more reasonable plot, the y-range should be specified. For example, to view the piece of the graph with $-20 \leq y \leq 20$, enter

> **plot((2*x^2-4)/(x+1),x=-6..6,y=-20..20);**

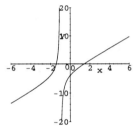

By changing the plot range, different aspects of the graph can be viewed. For example, changing the x- and y-ranges to **x=−50..50, y=−100..100** displays the graph for larger values of x, where the graph of the function approaches the line $y = 2x - 2$ as a skewed asymptote. However, with such large values of x, the vertical asymptote at $x = -1$ becomes obscured.

> **plot((2*x^2-4)/(x+1),x=-50..50,y=-100..100);**

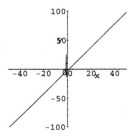

If the x-range is omitted from the plot command, then Maple will plot the expression over the interval $-10 \le x \le 10$ (in other words, this is the default range).

Note that the scale on the y-axis is different from the scale on the x-axis. Normally, Maple adjusts the scales of both axes so that the plot fills the plot window. The scale on the x- and y-axes can be forced to be the same by clicking on the word **Projection** on the menu bar in the plot window and then clicking on the **Constrained** menu option. An alternative is to use the modified command option **plot(expression,x,scaling=constrained);**. Other options to **plot** will be discussed in Section **2.4**.

More than one expression can be graphed on the same plot by enclosing several expressions using curly braces { }. For example, to plot the graphs of x^2 and x^3 on the same coordinate axis over the range $-2 \le x \le 3$, type

> **plot({x^2,x^3},x=-2..3);**

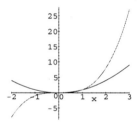

Caution: Opening too many plots on your computer will impede its performance. It is wise to close unneeded plots (be sure to close them rather than minimize them).

1.5   Summary

- Maple does everything that a graphing calculator will do.

- Maple function arguments must be in parentheses; for example, $\sin x$ must be written as **sin(x);**.

- Maple executes arithmetic commands in a predefined order. Use care in entering expressions (using parentheses) and examine output for correctness.

- Multiplication requires an asterisk. Juxtaposition of symbols is not allowed in Maple as a synonym for multiplication: **(x+4)(x+2)** is viewed as a function evaluation, not a product.

- Most calculators store numbers only as floating-point decimals, with a mantissa containing a preselected number of digits of accuracy and an exponent. Maple has an alternative exact mode: it treats the fraction 1/3 not as a decimal 0.3333333333 to any number of 3's, but instead as a list of two integers, 1 and 3.

- The distinction between storage modes is important because Maple will often not be able to express an answer in terms of a list of integers, and so it will parrot back the original expression typed in. This does not represent a syntax error: Maple simply doesn't know how to give an exact answer for **sin(1)**.

- Exact answers can be converted to floating-point decimal approximations of any order by using **evalf**.

- It is unnecessary to retype intermediate results in subsequent calculations. Maple provides two alternatives: the assignment command, :=, in which a label is associated with a given number output; and the percent, **%**.

- Once a number or algebraic expression is assigned a label, any statement that contains that label treats it as a synonym for the number or expression itself.

- Once an assignment has been made, Maple remembers that assignment until it is told otherwise. One common source of frustration is forgetting that a label already has an assigned value when trying to use it as a free variable.

- Labels with assigned values can be cleared by using single quotes.

- Algebraic expressions can be manipulated with the commands **expand**, **factor**, and **simplify**.

- Use **?command** to look up examples of proper command syntax. The command name does not have to be exact to get an answer. Also, use the Keyword Search facility in the **Help** window.

- Maple can be used to create graphs of expressions. Know how Maple produces graphs, the default domain and range, and the number of points plotted. Be aware that the defaults do not always show important features of the graph and know how to change the defaults to do so.

- Clicking the mouse cursor on a graph point shows its approximate coordinates.

1.6 Exercises

1. Assign the variable name a to the number $2\pi/5$ and then use **evalf** to compute the decimal approximations for a^2, $1/a$, $\sqrt{a}$, $a^{1.3}$, $\sin(a)$, and $\tan(a)$.

2. The number of significant digits can be changed from the default value of 10 to some other number, such as 20, with the command **Digits:=20;**. Repeat Exercise 1 with 20 significant digits. Alternatively, the **evalf** command can be modified. For example, to calculate a^2 to 20 significant digits, type **evalf(a^2,20);** . Return the number of digits to 10 by executing **Digits:=10;**.

3. Expand the following expressions.

   (a) $(x^2 + 2x - 1)^3(x^2 - 2)$

   (b) $(x + a)^5$

   Note: If you have done Exercise 1 or 2, the label a already has a value assigned to it. Recall that this value should be unassigned by typing **a:='a';**.

4. Factor the expression $x^2 + 3x + 2$. What happens if this expression is changed to $x^2 + 3.0x + 2.0$?

5. Try factoring $x^2 + 3x - 11$.

6. Factor $x^8 - 1$.

7. Simplify

$$\frac{2x^2}{x^3 - 1} + \frac{3x}{x^2 - 1}$$

8. Plot the graph of $y = \tan(x)$. Experiment with the x- and y-ranges to obtain a reasonable plot of one period of $y = \tan(x)$.

9. (a) Plot the graph of $y = \dfrac{3x^2 - 2x + 1}{x - 1}$ over a small interval containing $x = 1$, for example, $0 \leq x \leq 2$. Experiment with the y-range to obtain a reasonable plot. What happens to the graph near $x = 1$?

   (b) Now plot the same expression over a large interval such as $-100 \leq x \leq 100$. Note that the behavior of the graph near $x = 1$ is no longer apparent. Why do you think this happens?

10. Plot the expressions $\sin(x)$, $\sin(2x)$, and $\sin(4x)$ over the interval $0 \leq x \leq 2\pi$ on the same coordinate axes. Now plot the same expressions over the interval $0 \leq x \leq 4\pi$.

11. Compute 7! using Maple.

12. Use **evalf(Pi,40);** to give the first 40 digits of π.

13. Compute the exact and floating-point values of $\sin(\pi/4)$.

14. Compute the exact and floating-point values of $\sin(1)$.

15. Compute the number of seconds in one year, showing the units in your product as each factor is entered.

16. Just as **factor** will decompose a polynomial into irreducible factors, there is also an integer factor command that gives the prime decomposition of an integer. Use the integer factor command **ifactor** to show that $2^{23} - 1$ is not prime.

17. See what happens when the command **expand** is applied to $(a + b)/c$.

18. Use **expand** to change $\ln\left(\frac{ab}{c}\right)$. Practice the technique of entering the expression and checking its Maple output to see that it is entered correctly. Then go back and insert the **expand** command on the same line.

19. Factor the expression $e^{2x} - 1$, by first using **expand**, then **factor**.

20. Label the points $P_0 := [1, 2]$ and $P_1 := [3, 7]$ on the same input line. Load the student package by typing **with(student);**. Find out the syntax for determining the slope between the two points via **?slope**. Find the slope between P_0 and P_1 using the **slope** command. Assign the value that you find to the variable m.

21. Plot the expressions $-x/2 + 5/2$ and $-3x + 5$ on the same graph. Limit the plot to x-values between 0 and 2. Click on **Projection** and **Constrained** to avoid distortion in the plot. Click on the intersection with the mouse to find the coordinates of the intersection of the lines $y = -x/2 + 5/2$ and $y = -3x + 5$.

Chapter 2

Expressions, Functions, and Equations

Background Information: Read Chapter 1 in Stewart's **Calculus**.

This chapter introduces expressions, functions, equations and curve fitting in Maple. It also describes the syntax necessary to plot these objects, as well as the syntax involved with solving equations and fitting a curve to a given set of data points.

2.1   Expressions

In the previous chapter, we saw that we can assign labels to numbers (such as **Area:=25*Pi;**). In a similar way, labels can be assigned to expressions (which may contain variables). This is useful when repeatedly referring to a complicated expression. For example, suppose several operations are to be performed on the expression $2x^3 - 5x^2 + x + 2$. Instead of repeatedly typing this expression, it can be assigned to a variable name, such as f, and then referred to in the future by typing **f;**.

To assign the expression $2x^3 - 5x^2 + x + 2$ to the label f, enter

> **f:=2*x^3-5*x^2+x+2;**
$$f := 2x^3 - 5x^2 + x + 2$$

This is called a *Maple expression.* This expression can be factored.

> **factor(f);**
$$(x-1)(x-2)(2x+1)$$

To plot f over the interval $-2 \le x \le 3$, type

```
> plot(f,x=-2..3);
```

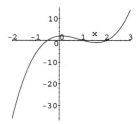

Expressions can involve more than one variable. For example, assign the variable vol the volume of a cylinder of radius r and height h.

```
> vol:=Pi*h*r^2;
```

$$vol := \pi\, h\, r^2$$

In this case, the volume involves two variables r and h, representing the radius and height of the cylinder.

There are two ways that values can be substituted into expressions. The first way is to assign values to the variables. For example, enter

```
> x:=6;
```

$$x := 6$$

The value $x = 6$ is automatically substituted into the expression f.

```
> f;
```

$$260$$

Now clear the value assigned to x by typing

```
> x:='x':
```

so that the label x is free for future use.

The second way to substitute values into variables in an expression is to use the **subs** command. For example, to substitute the value $x = 6$ into the expression $f = 2x^3 - 5x^2 + x + 2$, enter

```
> subs(x=6,f);
```
$$260$$

There is an important difference between the two methods. Using the first method, the value of f was changed to 260 (so if you try to plot f, for example, you get the horizontal line $y = 260$). With the second method, the value 260 is returned as output. However, the values of f and x have not been changed.

Other variables can be substituted for x. Substitute $x = a + h$.

```
> subs(x=a+h,f);
```
$$2(a+h)^3 - 5(a+h)^2 + a + h + 2$$

Again, the value of f has not been changed. To change the value of f, enter the command **f:=subs(x=a+h,f);**.

Multiple substitutions can be entered one at a time; if simultaneous substitution is desired, then curly braces { } may be used. For example, to substitute $r = 2$ and $h = 5$ into the expression for the volume of a cylinder, enter

```
> subs({r=2,h=5},vol);
```
$$20\,\pi$$

2.2 Functions

In the above discussion, the **subs** command is used to substitute values into expressions. For example, to substitute $x = 6$ into the expression $f := 2x^3 - 5x^2 + x + 2$, the command **subs(x=6,f);** is entered. However, in mathematics, the notation $f(6)$ is more commonly used to indicate the value of the function f at $x = 6$. This syntax will not make any sense in Maple unless f is first entered as a *function* instead of an *expression*. In this section, we describe how to enter functions in Maple using the arrow syntax.

To enter $f(x) = 2x^3 - 5x^2 + x + 2$ as a function, type

```
> f:=x->2*x^3-5*x^2+x+2;
```
$$f := x \rightarrow 2x^3 - 5x^2 + x + 2$$

The arrow symbol is entered by typing the *minus* key $-$ immediately followed by the *greater than* key $>$. This notation reflects the basic idea that a function is a rule (in this case f) that sends a number (labeled x) to ($\rightarrow$) an output value (in this case given by $2x^3 - 5x^2 + x + 2$).

The function f can be evaluated at a value, such as $x = 6$.

> f(6);

$$260$$

This function can also be evaluated at $x = a + h$.

> f(a+h);

$$2(a+h)^3 - 5(a+h)^2 + a + h + 2$$

Functions of two or more variables can also be entered in Maple. For example, to enter into Maple the function $vol(x, y) = x^2 y$ (which represents the volume of a rectangular box with square base of side length x and height y), type

> vol:=(x,y)->x^2*y;

$$vol := (x, y) \rightarrow x^2 y$$

To evaluate *vol* at $x = 3$ and $y = 5$, type **vol(3,5);**.

The syntax for many Maple commands depends on whether functions or expressions are involved. For example, to plot the function $f:=x \rightarrow 2x^3 - 5x^2 + x + 2$ over the interval $-2 \leq x \leq 3$, enter

> plot(f,-2..3);

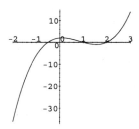

Note the difference between the syntax of this plot command and the syntax of the plot command if f had been defined as an expression. For an expression f, the syntax would be **plot(f,x=−2..3);**.

It is important to keep in mind the distinction between a function and an expression. If f is a function, then $f(x)$ is an expression in x. Therefore the plot command **plot(f(x),x=−2..3);** is also valid. However, a mix of syntax, such as **plot(f,x=−2..3);** or **plot(f(x),−2..3);** will result in error messages or empty plots.

Other commands can only be applied to an expression. For example, the command **factor(f);** will not work if f is defined as a function. However, $f(x)$ is an expression in x. Therefore the command **factor(f(x));** will work.

> **factor(f(x));**

$$(x-1)(x-2)(2x+1)$$

By contrast, we enter and factor the expression $g := 2x^3 - 5x^2 + x + 2$ as follows.

> **g:=2*x^3-5*x^2+x+2; factor(g);**

$$g := 2x^3 - 5x^2 + x + 2$$

$$(x-1)(x-2)(2x+1)$$

The following command reflects incorrect syntax.

> **factor(g(x));**

Because g has *not* been entered as a function, $g(x)$ does not make sense.

Since some commands only work with either a function or an expression, it is important to be able to change back and forth between them. To convert from a function to an expression is easy. As mentioned above, if f is entered as a function, then **f(x);** is the expression given by evaluating f at the variable x. Conversely, to convert from expressions to functions, there are two options. The first involves inserting the arrow notation **x ->** into the definition of f so that its definition coincides with the standard function syntax given above. The second involves the command **unapply**. We illustrate this command with an example. Suppose f is the expression

> **f:=2*x^3-5*x^2+x+2;**

$$f := 2x^3 - 5x^2 + x + 2$$

To turn f into a function, enter the command

> **f:=unapply(f,x);**

$$f := x \to 2x^3 - 5x^2 + x + 2$$

The second entry in **unapply(f,x);** tells Maple that the variable in the function will be x, since there may be other letters in the expression besides x. The first method involves searching for the defining statement for f, in order to insert the arrow. This is unnecessary when using **unapply.** Consequently, the command **unapply** can be very helpful if your Maple session is long and the defining statement for f is hard to find.

2.3   Solving Equations

Maple has two commands for solving equations, **solve** and **fsolve**, providing exact and approximate solutions, respectively. To solve $x^2 + 2x - 1 = 0$ enter

> **solve(x^2+2*x-1=0,x);**
$$-1 + \sqrt{2}, -1 - \sqrt{2}$$

Solutions can be given labels. The above solution can be labeled *sol.*

> **sol:=solve(x^2+2*x-1=0,x);**
$$sol := -1 + \sqrt{2}, -1 - \sqrt{2}$$

The first and second solutions can be referred to by typing **sol[1];** and **sol[2];**.

> **sol[1]; sol[2];**
$$-1 + \sqrt{2}$$

$$-1 - \sqrt{2}$$

The solution can now be checked with a plot of $f := x^2 + 2x - 1$ over an interval, such as $-3 \le x \le 3$, by entering **plot(f,x=-3..3);** to see that the graph of f crosses the x-axis at the solutions. Alternatively, the values $x = -1 + \sqrt{2}$ and $-1 - \sqrt{2}$ can be substituted into f (with the commands **subs(x=sol[1],f);** and **subs(x=sol[2],f);**) to see that the value of f is zero at these solutions.

Labels can be given to part of an equation or to an entire equation. For example, the left side of the equation in the **solve** command can be labeled f

> **f:=x^2+2*x-1;**
$$f := x^2 + 2x - 1$$

and then the above equation can be solved with the command

> **sol:=solve(f=0,x);**
$$sol := -1 + \sqrt{2}, -1 - \sqrt{2}$$

Another approach is to assign a variable name to the entire equation. To assign the name *eq* to the above equation, enter

> `> eq:=x^2+2*x-1=0;`

$$eq := x^2 + 2\,x - 1 = 0$$

This equation can now be solved.

> `> solve(eq,x);`

$$-1 + \sqrt{2}, -1 - \sqrt{2}$$

Labeling either the equation or the left side of the equation is useful because the definition of the label can be checked to see that the equation is entered correctly before Maple solves it.

More than one equation can be solved by using curly braces { }. Consider the following two linear equations.

> `> eq1:=3*x+2*y=1; eq2:=x+2*y=3;`

$$eq1 := 3\,x + 2\,y = 1$$

$$eq2 := x + 2\,y = 3$$

Both of these equations can be solved simultaneously.

> `> solve({eq1,eq2},{x,y});`

$$\{\,x = -1, y = 2\,\}$$

Although this system of equations has been solved, the values of x and y have not been changed (even though the screen displays $x = -1$ and $y = 2$). To assign these values to x and y, use the **assign** command.

> `> assign(%); x, y;`

$$-1, 2$$

Now clear the values from x and y (**x:='x'; y:='y';**).

Exact solutions to complicated equations can be difficult or impossible to find. For example, enter the equation

```
> eq:=x^3-2*x^2+x-3=0;
```
$$eq := x^3 - 2\,x^2 + x - 3 = 0$$

Solving this equation with the command **solve(eq,x);** will yield three solutions that are very complicated (try this). The solutions are so complicated that, to conserve space on the screen, the answers are abbreviated with the label *%1*, whose value is given at the end of the output. To get decimal approximations to these solutions, change the coefficients of the equation from exact integers to floating-point decimals.

```
> eq:=x^3-2.0*x^2+x-3.0=0;
```
$$eq := x^3 - 2.0\,x^2 + x - 3.0 = 0$$

Now solving the equation will yield nicer (but only approximate) decimal answers.

```
> solve(eq,x);
```

$$2.174559410,\ -.0872797049 + 1.171312111\,I,$$
$$-.0872797049 - 1.171312111\,I$$

Note that the answers involving $I = \sqrt{-1}$ are complex numbers.

Some equations are impossible to solve exactly. In fact, there is a famous theorem in mathematics that states that there is no formula (analogous to the quadratic formula) for finding roots of polynomials of fifth degree or higher. Consider the following equation.

```
> eq:=x^7+3*x^4+2*x-1=0;
```
$$eq := x^7 + 3\,x^4 + 2\,x - 1 = 0$$

Try finding the exact solutions using Maple.

```
> solve(eq,x);
```
$$\mathrm{RootOf}(\,\_Z^7 + 3\,\_Z^4 + 2\,\_Z - 1\,)$$

Maple's response indicates that it does not know how to solve this equation exactly. In this situation, use Maple's **fsolve** or **allvalues** commands, which will find approximate solutions.

```
> fsolve(eq,x);
```

$$.4414177090$$

The **fsolve** command is best used in conjuction with a plot. Consider the equation

```
> eq:=x^2+1/x-1/x^2=0;
```

$$eq := x^2 + \frac{1}{x} - \frac{1}{x^2} = 0$$

Solving this equation with **fsolve** will yield one solution

```
> fsolve(eq,x);
```

$$-1.220744085$$

However, a plot of the expression $f := x^2 + 1/x + 1/x^2$ indicates that there is another solution.

```
> plot(x^2+1/x-1/x^2,x=-2..2,y=-20..20);
```

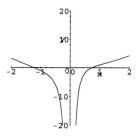

This second solution can be found with the following modification of the **fsolve** command:

```
> fsolve(eq,x,x=0..1);
```

$$.7244919590$$

The **fsolve** command will find all the real roots of a polynomial, but as the previous example shows, more complicated equations should be handled by a plot together with specifying a range with the **fsolve** command.

Example. Find the equation of the parabola that passes through the points $(-1, 2)$, $(1, -1.5)$, and $(4, 7)$.

Solution. First, enter the general equation of a parabola (defined as a function of x).

```
> p:=x->a*x^2+b*x+c;
```
$$p := x \rightarrow a\,x^2 + b\,x + c$$

The unknown coefficients a, b, and c must be found so that the parabola passes through the points $(-1, 2)$, $(1, -1.5)$, and $(4, 7)$. In order for the parabola to pass through the point $(-1, 2)$, the equation $p(-1) = 2$ must be satisfied. We enter this equation with the label *eq1*.

```
> eq1:=p(-1)=2;
```
$$eq1 := a - b + c = 2$$

The other two equations corresponding to the other two points are entered similarly.

```
> eq2:=p(1)=-1.5; eq3:=p(4)=7;
```
$$eq2 := a + b + c = -1.5$$

$$eq3 := 16\,a + 4\,b + c = 7$$

These equations are then solved and the label *sol* is assigned to the solution.

```
> sol:=solve({eq1,eq2,eq3},{a,b,c});
```
$$sol := \{\, a = .9166666667, c = -.6666666667, b = -1.750000000 \,\}$$

(The order of your solutions may differ.) These values of a, b, and c can be substituted into $p(x)$ using the **subs** command.

```
> f:=subs(sol,p(x));
```
$$f := .9166666667\,x^2 - 1.750000000\,x - .6666666667$$

Here the label f is given to the parabola. Note that f is defined as a *Maple expression*, which can be plotted with the command **plot(f,x=-2..5);** to see if the parabola passes through the desired points (try this).

Maple can be used to plot an equation, but the **implicitplot** command must be used instead of **plot**. In order to use this command, first load the **plots** package with the command

```
> with(plots);
```

Maple responds with a list of special plot commands, one of which is **implicitplot**.

Enter the equation, labeled *eq*, followed by the **implicitplot** command.

> eq:=x^2+y^2=1; implicitplot(eq,x=-1..1,y=-1..1);
$$eq := x^2 + y^2 = 1$$

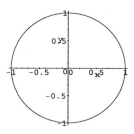

Unlike the **plot** command, the **implicitplot** command does not have a default range. The x- and y-ranges must be entered each time.

2.4 More on Plots

In this section we discuss the following options to the **plot** and **implicitplot** commands.

plot(f,x=a..b,style=point);	A data-point plot of the expression f.
plot(f,x=a..b,title='plot title');	For specifying title in the plot.
plot(f,x=a..b,numpoints=n);	Changes the resolution (the number of data points) in the plot.
implicitplot(eq,x≐a..b,y=c..d,grid=[m,n]);	Changes the resolution of the implicitplot of the equation *eq*. Use values of $m \le 70$ and $n \le 70$.
plot3d(f,x=a..b,y=c..d);	A three-dimensional plot of the expression f.
plot([[1,2],[-2,3],[4,4]],style=point);	Plots a list of points.
plot([[1,2],[-2,3],[4,4]],style=line);	Plots a list of points with connecting line segments.
with(plots); display([g1,g2]);	Displays two graphs $g1$ and $g2$ (see below).
plot(f,x,discont=true);	Plots a discontinuous expression f.
plot([x(t),y(t),t=a..b]);	Plots the parametric curve $x(t)$ and $y(t)$ for t in $[\,a,b]$

Point Plots

Sometimes, it is desirable to obtain a point plot of a graph of a function or an expression (instead of a continuous curve). This is especially true when the function has a discontinuity.

Example. Plot the graph of the expression $f := \dfrac{x - 0.21}{|x - 0.21|}$ over the interval $-2 \le x \le 2$. This expression is discontinuous (i.e., there is a break in the graph) at $x = 0.21$. However, a plot of this expression with the command **plot(f,x=-2..2);** joins together the two disconnected pieces of this graph with a connecting line segment. A more realistic plot can be obtained using the **style=point** option.

> f:=(x-0.21)/abs(x-0.21); plot(f,x=-2..2,style=point);
$$f := \frac{x - .21}{|x - .21|}$$

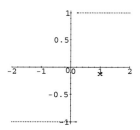

An alternative is to use your mouse to click on the menu option labeled **Style** at the top of the plot window and then click on the **Point** option. There is also a **discont=true** option to the **plot** command for plotting discontinuous expressions.

Change in Resolution.

The number of points in this plot can be increased from the default setting (of around 49) to another number, say 70, by adding the **numpoints** option.

> plot(f,x=-2..2,style=point,numpoints=70);

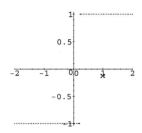

The **numpoints** option can be added to continuous plots to increase the resolution of the graph (see Exercise 19). Do not use too large a value of **numpoints.** Otherwise the computer can become overloaded (say, **numpoints** < 500).

In a similar manner, the resolution of the **implicitplot** command can be increased by adding the option **grid=[m,n],** where m specifies the number of grid points along the x-axis and n specifies the number of grid points along the y-axis. The default value of m and n is 25. Keep the values of m and n from getting too large so that the computer does not get overloaded. A safe limit for m and n is 70.

Three-Dimensional Plots

We have discussed expressions and functions of two or more variables. A plot of a function or an expression of two variables can be obtained by using the **plot3d** command.

Example. Plot the expression **f:=sin(2*x+y);** over the region $-5 \leq x \leq 5, -5 \leq y \leq 5$.

Solution. First enter this expression in Maple. Then enter a **plot3d** command.

```
> f:=sin(2*x+y); plot3d(f,x=-5..5,y=-5..5,style=patch);
```
$$f := \sin(2x + y)$$

As with **implicitplot**, the resolution of this graph can be increased by adding the option **grid=[50,50]**. To rotate the graph in space, refer to **Help** in the **3D** plot window.

If f is entered as a function rather than as an expression, then the **plot3d** command can be entered as

```
> f:=(x,y)->sin(2*x+y); plot3d(f(x,y),x=-5..5,y=-5..5);
```

or

```
> plot3d(f,-5..5,-5..5);
```

Plotting a List

Often, problems involve data rather than functions or expressions. For example, to plot the data set [[1,2], [-2,3], [4,4]], first assign it to a variable, say *mylist*

```
> mylist:=[[1,2], [-2,3], [4,4]];
```
$$mylist := [[1,2],[-2,3],[4,4]]$$

and check for typing errors. Then issue the command

```
> plot(mylist,style=point);
```

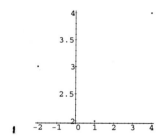

Changing the option to **style=line** will draw this data set with connecting line segments.

Displaying Several Plots

As mentioned in Chapter 1, a plot of two or more expressions or functions over the same interval can be obtained by enclosing the expressions with curly braces { }. To obtain a plot of two or more expressions over *different* intervals, the **display** command must be used.

Example. Plot the expression $f := x^2$ over the interval $-1 \leq x \leq 1$ and the expression $g := 2x - 3$ over the interval $0 \leq x \leq 2$.

Solution. First define these expressions in Maple, then assign their individual plots to variable names, say *p1* and *p2*, ending with colons rather than semicolons.

```
> f:=x^2; g:=2*x-3; p1:=plot(f,x=-1..1): p2:=plot(g,x=0..2):
```

$$f := x^2$$

$$g := 2\,x - 3$$

The colons are used to suppress output (the use of semicolons would display lengthy plot data structures). Next load the **plots** package by typing **with(plots);**. Then enter the **display** command (this time with a semicolon).

```
> display([p1,p2]);
```

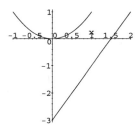

2.5   Curve Fitting

It is a simple matter to use Maple to fit a curve to a set of data points. As an illustration, consider the set of data points given in the table below:

x	1	2	3	4	5	6	7
y	2.2	3.7	5.5	5.9	5.6	4.2	3.1

We can obtain a plot of these points in Maple by using the **plot** command (see Section 2.4). We demonstrate a slightly different approach below. Simply load the **stats** package and utilizing the **scatter2d** command.

```
> with(stats):
> xvals:=[1,2,3,4,5,6,7];
> yvals:=[2.2,3.7,5.5,5.9,5.6,4.2,3.1];
```

$$xvals := [1, 2, 3, 4, 5, 6, 7]$$

$$yvals := [2.2, 3.7, 5.5, 5.9, 5.6, 4.2, 3.1]$$

```
> with(statplots):
> scatter2d(xvals,yvals);
```

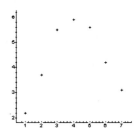

Although this method of plotting points might seem more complicated, it has some advantages. For example, Maple's built-in curve fitting commands require the x and y coordinates to be entered separately.

From the plot, it appears as though a quadratic function might give a good approximation to this data. The following Maple commands will create a quadratic fit.

> **fit[leastsquare[[x,y],y=z*x^2+b*x+c]]([xvals,yvals]);**
$$y = -.3619047619\,x^2 + 3.030952381\,x - .5714285714$$

> **f:=rhs(%);**
$$f := -.3619047619\,x^2 + 3.030952381\,x - .5714285714$$

Let's see how well we did.

> **with(plots):**
> **p1:=scatter2d(xvals,yvals):**
> **p2:=plot(f,x=0..8):**
> **display({p1,p2});**

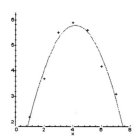

Note that the **fit[leastsquare[...]]**command can be easily modified for other types of curve fits. For example, consider the population data given in the table below.

time (decades since 1900)	0	1	2	3	4	5
Population (billions)	1.65	1.75	1.86	2.07	2.30	2.52

6	7	8	9	9.6
3.02	3.70	4.45	5.30	5.77

The process of entering this data into a Maple session and obtaining a cubic fit is shown below.

```
> tvals:=[0,1,2,3,4,5,6,7,8,9,9.6]:
> Pvals:=[1.65,1.75,1.86,2.07,2.30,2.52,3.02,3.70,4.45,5.30,5.77]:
> p1:=scatter2d(tvals,Pvals):
> fit[leastsquare[[t,P],P=a*t^3+b*t^2+c*t+d]]([tvals,Pvals])
```

$$P = .002325670340\,t^3 + .01914440256\,t^2 + .03583942511\,t + 1.676870710$$

```
> f:=rhs(%);
```

$$f := .002325670340\,t^3 + .01914440256\,t^2 + .03583942511\,t + 1.676870710$$

```
> p2:=plot(f,t=0..10):
> display({p1,p2});
```

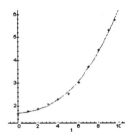

2.6 Summary

- Distinguish between Maple expressions, Maple functions, and Maple equations.

- Know how to form a Maple expression and how to use arrow notation to form a Maple function.

- Convert from a Maple expression to a Maple function either by inserting arrow notation or by using the **unapply** command. An evaluated Maple function becomes a Maple expression.

- Know the proper syntax to substitute a value into either a Maple expression or a Maple function.

- Distinguish: the assignment command globally changes values; **subs** affects only screen output from a given expression. Know how to combine the two to save the screen output for later use.

- Distinguish the syntax for plotting a Maple expression and that for plotting a Maple function. Recognize that neither form is appropriate for plotting a Maple equation.

- Use the **implicitplot** command in the **with(plots)** library to graph Maple equations.

- Distinguish: **evalf** converts an exact number to floating-point decimal format; **solve** and **fsolve** solve equations.

- Distinguish: **solve** attempts to produce exact answers if the input equation contains exact values; **fsolve** produces floating-point decimals.

- Know when to use **solve** (linear equations, including systems; quadratics with exact coefficients; equations where the answer will contain variables) and when to use **fsolve** (when you only want floating-point decimal answers; where **solve** produces **RootOf**; most other cases).

- Know that except in the case of polynomials **fsolve** may not find all solutions. Help it find a desired root by using a plot and clicking to determine a restricted search interval.

- Be able to display two graphs simultaneously on the same coordinate axes.

- Click the mouse in the Plot window menu or use **plot** command options to adjust the display of plots.

- Be able to fit various curves to given sets of data points.

- Be able to plot data points and curves on the same graph.

2.7   Exercises

1. Enter an expression that describes the area of a square in terms of its diagonal length. Use the variable name *area* for the area of the square and let d be the diagonal length. Use the **subs** command to compute the area of the square when $d = 2$ and $d = 3.7$.

2. Enter an expression that describes the area of a circle in terms of its circumference. Use the variable name *area* for the area of the circle and let p be the circumference. Use the **subs** command to compute the area of the circle when $p = 2$ and $p = 3.7$.

3. Enter an expression that describes the volume of a cube in terms of its diagonal length. Use the variable name *vol* for the volume of the cube and let d be the diagonal length. (If you have done Exercise 1, be careful to unassign the variable d before you enter this expression.) Use the **subs** command to compute the volume of the cube when $d = 2$ and $d = 3.7$.

4. Repeat Exercises 1–3 with the expressions entered as Maple functions. Evaluate these functions when the independent variable (either d or p) is 2 and 3.7.

5. (a) A large tank initially contains 100 gallons of salt water with a concentration of 1/10 lb of salt per gallon. Salt water with a concentration of 1/2 lb per gallon flows into the tank at the rate of 3 gallons per minute.

   (b) Express the total amount of salt t minutes later as an expression in t; label this amount S. Use the **subs** command to evaluate S at $t = 4, 10$, and 50 minutes.

   (c) Express the concentration as an expression of t and evaluate the concentration at $t = 4, 10$, and 50 minutes.

6. (a) Enter the function that describes the volume of a cylinder of radius r and height h as a function of the two variables r and h. Evaluate this function at $r = 2$ and $h = 3$.

   (b) Compute an expression for the volume of a prism with a height h and with base given by an equilateral triangle of side length s. Enter this expression as a function of the two variables s and h. Compute the volume of the prism when $s = 2.3, h = 4.5$ and when $s = 1.7, h = 7.2$.

7. Enter $f = \dfrac{x^5 + 1}{x^2 - 1}$ as an expression. Plot the expression f over the interval $-2 \le x \le 0$. Evaluate the limit $\lim\limits_{x \to -1} f$.

8. Enter $f(x) = \dfrac{x^4 - 16}{x^2 - 4}$ as a function. Plot the graph of $f(x)$ over the interval $1 \le x \le 3$. Evaluate the limit $\lim\limits_{x \to 2} f(x)$.

9. Consider the expression entered in Maple as **f:=x ^2-1;**. Which of the following statements involve incorrect Maple syntax?

   (a) **plot(f,x=−2..2);**
   (b) **plot(f(x),x=−2..2);**
   (c) **factor(f);**
   (d) **subs(x=2,f);**
   (e) **subs(x=2,f(x));**
   (f) **f(2);**
   (g) **solve(f=0,x);**
   (h) **solve(f(x)=0,x);**

10. Repeat Exercise 9 if f is now entered as a *function* **f:=x->x ^2-1;**.

11. Enter each of the following as an expression: $f = (x + 2)/(2x + 1)$ and $g = x/(x - 2)$. Convert each expression to a function using **unapply**. Find the composition of the two functions by $f(g(x))$ and by $(f@g)(x)$. Here, the @ symbol is used to indicate the composition, just as $\circ$ is used in standard mathematical notation. Note that the output is an expression, since each function has been evaluated.

12. Use the **plot** command to draw the graph of $f = \dfrac{\sqrt{4 - x} + \sqrt{3 + x}}{x^2 - 2}$ showing only $-5 \le x \le 5$ and $-10 \le y \le 10$. Notice the vertical line segments that Maple puts in. Click on **Style** and **Point** on the graph window to show only the points that Maple used. Verify that the lines are not really there. Change the expression f into a function by using **f:=unapply(f,x);**. Plot the function f, showing only the range of x and y indicated above. Find $f(-1), f(2), f(5)$. Reconvert f to an expression and use **subs** to find $f(-1), f(2), f(5)$.

13. Form the expression $f := \sqrt{(1/3)x^2 \sin(x + \pi/6)}$. Plot this expression and click on the endpoints of the pieces to determine the domain of the expression. (You may find that the labeling on the axes gets in the way. Click on **Axes** and **None** to remove them.) Use **fsolve** on $\sin(x + \pi/6) = 0$, with the information that you obtained from the plot to limit the range, and determine to ten decimal places the endpoints of the first connected segment of the domain that lies to the right of $x = 0$.

14. Solve the following equations using the **solve** command. Check your answers with a plot. In each case substitute the roots into the expression on the left side of the equation to verify that the roots satisfy the equation.

    (a) $x^2 + 3x + 1 = 0$

    (b) $x^2 + 3.0x + 1.0 = 0$

    (c) $x^3 + x + 1 = 0$

    (d) $x^3 + x + 1.0 = 0$

    (e) $x^2 + 2x + 2 = 0$

    (f) $x^2 + 2.0x + 2.0 = 0$

15. Use the **plot** and **fsolve** commands to find all solutions to the following equations over the given interval.

    (a) $\sin^2 x = \cos(3x), 0 \le x \le \pi$

    (b) $\cos^2 x = \sin(3x), 0 \le x \le \pi$

    (c) $8 \cos x = x, -\infty < x < \infty$

16. Solve the system of equations $3x + y = 2$ and $2x - 3y = 7$ for x and y.

17. Find the decimal approximations for all roots of the equation

$$\frac{1}{x^4} - 3 + x^2 = 0$$

(make sure you plot this one).

18. Find the equation of the cubic that passes through the points $(1, 2.4)$, $(3, 5.6)$, $(4, -2.7)$, and $(7, 4.7)$. Graph your answer and see if it passes through these points.

19. Define the expression **f:=x*sin(1/x);** in Maple. Enter the following plot commands and examine the behavior of each plot near the origin.

 (a) **plot(f,x=-1..1);**

 (b) **plot(f,x=-1..1,numpoints=100);**

 (c) **plot(f,x=-1..1,numpoints=300);**

20. Use the **plot3d** command to view the graphs of the following expressions:

 (a) $f := x^2 - y^2$

 (b) $f := x^2 + y^2$

 (c) $f := x^4 - y^4$

 (d) $f := (x^2 - y^2)(x + y)$

 (e) $f := \cos(x + y)$

21. For each of the following, plot the pair of expressions over the given intervals using **display** (see Section 2.4).

 (a) $f := x^4 - 2x^2, -2 \le x \le 2$, and $g := x^3 - x, -1 \le x \le 3$.

 (b) $f := \sin(x) + \cos(x), -2\pi \le x \le 0$, and

$$g := \frac{\tan(x)}{\tan(x) + 1}, 0 \le x \le \pi/2$$

22. The general equation of a circle is

$$x^2 + y^2 + ax + by = c$$

Find the equation of the circle (i.e., find a, b, and c) that passes through the three points $(1, 1)$, $(3, -1)$, and $(6, 4)$.

Hint: It is not possible to graph the circle with a function of the form $y = f(x)$ (why?); so it is easier to work with expressions using the **subs** command to obtain the necessary three equations. Start by assigning the above equation to the variable name *eq*. Since the circle is to contain the point $(1, 1)$, the first equation is obtained from the command

eql:=subs({x=1,y=1},eq);. The other two equations are obtained in an analogous fashion using the other two points. Then use the **solve** command to solve these three equations for a, b, and c. After substituting these values for a, b, and c into the equation of the circle, plot your answer with the **implicitplot** command. Type **with(plots);** and then **implicitplot(eq,x=-10..10,y=-10..10);.**

23. The point of this exercise is that plots can be deceiving. Enter the expression $f := x^3 - x^2 - x + 1.001$. Plot f over the interval $-2 \leq x \leq 2$. From this plot, guess how many real solutions there are to the equation $f = 0$. Now solve the equation $f = 0$ with the **solve** command. How many solutions did **solve** return? Now replot f over the interval $0.9 \leq x \leq 1.1$. Does the graph of f cross the x-axis near the point $x = 1$?

24. The point of this exercise is to show how piecewise defined functions can be entered into Maple. Consider the function

$$f(x) = \begin{cases} x^2 & \text{if } x \leq 1 \\ 2x + 1 & \text{if } x > 1 \end{cases}$$

This function can be entered into Maple with the following **if then else fi** syntax.

```
> f:=x->if x <= 1 then x^2 else 2*x+1 fi;
```

Note that **fi** terminates the **if then** construct. The function f can now be evaluated like any other function. For example, the value of f at $x = 0.5$ and $x = 2$ can be obtained by entering **f(0.5);** and **f(2);**. To plot f over the interval $-1 \leq x \leq 3$ enter

```
> plot(f,-1..3);
```

Using $f(x)$ in place of f in this command would result in the error message `cannot evaluate Boolean`. As mentioned in Section 2.4, Maple plots functions and expressions by plotting points and then connecting the dots with line segments. So the jump discontinuity of f at $x = 1$ is not apparent. The discontinuity at $x = 1$ becomes apparent with the **style=point** option.

```
> plot(f,-1..3,style=point);
```

Functions with more than two pieces can be defined in Maple as well. For example, the function

$$f(x) = \begin{cases} x^2 & \text{if } x \leq 0 \\ 2x + 1 & \text{if } 0 < x \leq 1 \\ -2x & \text{if } x > 1 \end{cases}$$

can be entered as

> f:=x->if x<=0 then x^2 elif x<=1 then 2*x+1 else -2*x fi;

The syntax **elif** stands for "else if." As many **elif** statements can be inserted as desired.

Try entering and plotting the following functions

$$f(x) = \begin{cases} \sin(x) & \text{if } x < 0 \\ \cos(x) & \text{if } x \geq 0 \end{cases}$$

$$g(x) = \begin{cases} 2x & \text{if } x \leq 0 \\ \sqrt{x} & \text{if } 0 < x \leq 2 \\ x^2 & \text{if } x > 2 \end{cases}$$

Also read about the new **piecewise** command via the Maple Help Facility.

25. *Draw a map of Texas.* To plot a map of Texas, enter the following two lists (named *north* and *south* for the northern boundary and southern boundary of Texas).

> north:=[[0,0],[3,0],[3,4.5],[6,4.5],[6,2.2],[7,2.1],[8,1.8],
> [9,1.9],[10,1.8],[11,1.7],[11,-2.2]];
> south:=[[0,0],[1,-1.1],[2,-2.5],[3,-2.9],[4,-2.3],[5,-2.8],
> [6,-4.4],[7,-5.8],[8,-6.1],[9,-3.3],[10,-2.8],[11,-2.2]];

Here, the origin is the western corner of Texas (near El Paso) and the x-axis is the extension of the east-west border between New Mexico and Texas. Each unit represents approximately 69 miles. After entering these lists, type **plot({north,south},style=line);** as done in the example in the text.

26. Use Maple to find the equation of the circle with center at $x = 3$ and $y = 3$ with radius 2. Then graph the circle in two ways:

(a) **solve** the equation for y and label the solution. Use each expression in the solution. The two expressions can be placed in curly braces to plot simultaneously. (Note that the circle looks more like an ellipse than a circle. What can you do to remove the distortion?)

(b) Remind Maple about the **with(plots):** command and do an **implicitplot** of the equation.

Remember to close your plots after you are finished.

27. Suppose a rubber band is stretched and the following data is recorded relating the restoring force to displacement.

Displacement (meters)	.01	.02	.03	.05	.06	.08	.10
Force (Newtons)	.21	.42	.63	.83	1.0	1.3	1.5

.13	.16	.18	.21	.25	.28
1.7	1.9	2.1	2.3	2.5	2.7

Determine whether this data is best approximated using a linear, quadratic or logarithmic curve fit (the logarithmic fit should be of the form $y = a + b\ln(x+1)$). In each case, plot the data against the curve.

28. The following table lists the temperature vs specific heat of air at low temperatures.

Temperature (K)	300	400	500	600
Specific Heat (J/gm-K)	1.0045	1.0134	1.0296	1.0507

700	800	900
1.0743	1.0984	1.1212

Determine a linear curve that approximates this data, and plot the curve and the data in the same coordinate system. In addition, compute the predicted values for specific heat (given by the linear approximation) corresponding to the temperature values in the table. Is there a relationship between the average of the predicted values and the average of the specific heat values in the table?

29. The following table lists the temperature and specific heart of air at high temperatures.

Temperature (K)	1000	1500	2000	3000
Specific Heat (J/gm-K)	1.1410	1.2095	1.2520	1.2955

Determine a quadratic curve that approximates this data, and plot the curve and the data in the same coordinate system. In addition, compute the predicted values for specific heat (given by the quadratic approximation) corresponding to the temperature values in the table. Is there a relationship between the average of the predicted values and the average of the specific heat values in the table?

30. The following Maple commands make it possible to easily generate random data sequences of integers between -10 and 10.

```
> with(stats):
> r:=rand(-10..10):
```

For example, if you want to create random data sequences named *xvals* and *yvals* containing 10 data values each, simply do the following:

```
> xvals:=[seq(r(),i=1..10)];
> yvals:=[seq(r(),i=1..10)];
```

$$xvals := [-4, 7, 8, 10, -6, -8, -5, 7, 6, -6]$$

$$yvals := [0, 5, -10, 1, 1, -3, -10, 5, -4, -8]$$

Note that the values obtained might be different than the ones shown above. Use Maple to create a random pair of data sequences named *xvals* and *yvals* containing 10 data values each. Give a linear curve that approximates this data and compute the predicted y values (given by the linear approximation) for the corresponding listed x values in *xvals*. Compare the average of the predicted y values to the given values in *yvals*. Repeat this process 5 times. What do you conclude?

Chapter 3

Differentiation

Background Information: Read Sections 2.1 and 2.6, and Chapter 3 in Stewart's **Calculus**.

We start this chapter by computing derivatives using the limit definition of the derivative. Then we introduce the Maple syntax for differentiation. Maple makes a distinction between expressions and functions (see Chapter 2), and we discuss the syntax for differentiating both. We conclude this chapter with sections on implicit differentiation and linear approximation.

3.1   The Limit of the Difference Quotient

Background Information: Read Sections 2.6 and 3.1 in Stewart's **Calculus**.

The definition of the derivative of the function $f(x)$ at the point $x = a$ is given by

$$f'(a) = \lim_{h \to 0} \frac{f(a+h) - f(a)}{h}$$

The motivation of this definition is that the derivative of f at $x = a$ should be the slope of the tangent line to the graph of f at $x = a$. The key idea is that the quantity $\frac{f(a+h)-f(a)}{h}$ is the slope of the line segment connecting the points $(a, f(a))$ and $(a + h, f(a + h))$. The definition of the derivative given above reflects the fact that the slope of the tangent line at $x = a$ is the limit of the slopes of these line segments as the number h tends to zero.

As an example, consider the function $f := x \to x^3 - 8x$ (enter this as a function in Maple). To compute $f'(2)$, calculate the above limit with $a = 2$.

```
> f:=x->x^3-8*x; Limit((f(2+h)-f(2))/h,h=0); value(%);
```

$$f := x \rightarrow x^3 - 8x$$

$$\lim_{h \to 0} \frac{(2+h)^3 - 8 - 8h}{h}$$

$$4$$

Note that the **Limit** command displays the limit so that you can check for typing errors and then the **value(")** command evaluates this limit (recall that the percent **%** refers to the result of the previous command, which in this case is the limit).

The value of $f'(2)$ is 4. This is the slope of the tangent line to f at $x = 2$. The tangent line also passes through the point $(2, f(2)) = (2, -8)$. Accordingly, its formula is

```
> y:=4*(x-2)-8;
```

$$y := 4x - 16$$

Both the tangent line and the function can be plotted on the same coordinate axes.

```
> plot({y,f(x)},x=0..4);
```

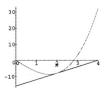

To see that this tangent line closely approximates the graph of the function f near $x = 2$, replot this graph over a small interval about $x = 2$.

```
> plot({y,f(x)},x=1.8..2.2);
```

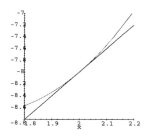

Another way to obtain $f'(2)$ is to compute $f'(x)$ and then substitute $x = 2$. To do this, compute the above limit with $a = x$ and store the result in the variable named *Df*.

> **Limit((f(x+h)-f(x))/h,h=0); Df:=value(%);**

$$\lim_{h\to 0} \frac{(x+h)^3 - 8h - x^3}{h}$$

$$Df := 3x^2 - 8$$

The derivative of f at a general point x is $Df = 3x^2 - 8$. To obtain $f'(2)$, substitute $x = 2$ into *Df* with the command **subs(x=2,Df);** and obtain the output 4 as above.

Let's examine the above limit process more carefully. First, enter the difference quotient $\dfrac{f(x+h) - f(x)}{h}$ into Maple and store this expression as the variable *diffq*.

> **diffq:=(f(x+h)-f(x))/h;**

$$diffq := \frac{(x+h)^3 - 8h - x^3}{h}$$

Next, simplify this expression.

> **simplify(diffq);**

$$3x^2 + 3xh + h^2 - 8$$

Now note that the derivative, $3x^2 - 8$, is obtained by letting h tend to zero (any term containing an h will disappear).

3.2   Differentiating Functions

Background Information: Read Sections 3.1 - 3.5 in Stewart's **Calculus.**

All the rules for differentiation are programmed into Maple, making it easy to differentiate complicated functions and expressions. The syntax for differentiating functions is as follows.

Suppose that f is defined as a Maple function. Then **D(f)** is the *function* that represents the derivative of f.

Example 1. Enter $f(x) = x^2(x^5 + 1)$ as a function.

> ```
> f:=x->x^2*(x^5+1);
> ```
$$f := x \to x^2\,(x^5 + 1)$$

This function can be differentiated with the command

> ```
> D(f);
> ```
$$x \to 2\,x\,(x^5 + 1) + 5\,x^6$$

The arrow notation indicates that **D(f)** is a function. To evaluate $f'(2)$, type

> ```
> D(f)(2);
> ```
$$452$$

Try evaluating f' at the points $x = -3$, $x = 3$, and $x = t$, by typing **D(f)(-3); D(f)(3);** and **D(f)(t);** respectively.

Example 2. Compute the equation of the tangent line at $x = 2$ of the function

> ```
> f:=x->(x^3-1)/(x+2);
> ```
$$f := x \to \frac{x^3 - 1}{x + 2}$$

The tangent line passes through the point $(2, f(2))$ and has a slope equal to $f'(2)$, which in Maple is **D(f)(2)**. First assign the slope to the variable m.

> ```
> m:=D(f)(2);
> ```
$$m := \frac{41}{16}$$

The formula of the tangent line is now given by the expression

```
> y:=m*(x-2)+f(2);
```
$$y := \frac{41}{16}x - \frac{27}{8}$$

A plot of the expression $f(x)$ and the expression y will verify that the correct tangent line has been found (enter **plot({f(x),y},x=0..4);**).

Higher derivatives can be evaluated by repeated application of the **D** operator. For example, **D(D(f));** represents the second derivative of the function f (so $f''(2)$ can be computed by entering **D(D(f))(2);**). Alternatively, the syntax **(D@@2)(f);** also represents the second derivative. This syntax is preferred for higher derivatives; e.g., $f'''(2)$ is evaluated by typing **(D@@3)(f)(2);**.

3.3 Differentiating Expressions

Background Information: Read Sections 3.1 - 3.5 in Stewart's **Calculus**.

Sometimes, it is more convenient to deal with expressions rather than functions. The **D** syntax, however, only works for functions. To differentiate an expression, the syntax is

diff(expression,x);

The above syntax assumes that the expression is to be differentiated with respect to x.

Example 1. Differentiate $3x^2 + 4x^3$ with respect to x.

```
> diff(3*x^2+4*x^3,x);
```
$$6x + 12x^2$$

Alternatively, this expression can be assigned to the variable f to make sure the expression f is entered correctly before differentiating it.

```
> f:=3*x^2+4*x^3;
```
$$f := 3x^2 + 4x^3$$

```
> Df:=diff(f,x);
```
$$Df := 6x + 12x^2$$

Here, the derivative is assigned to the variable **Df**. To evaluate the derivative at a point, say $x = 2$, the **subs** command must be used.

```
> subs(x=2,Df);
```
$$60$$

Example 2. Enter the expression

```
> f:=cos(t^2)^2;
```
$$f := \cos(t^2)^2$$

Now differentiate f with respect to t.

```
> diff(f,t);
```
$$-4\cos(t^2)\sin(t^2)\,t$$

Higher derivatives can be calculated by using a dollar sign followed by the number of derivatives that is to be taken. For example, the second derivative of f can be computed by entering the command **diff(f,t\$2);** (try this).

If f is a function, then $f(x)$ is an expression in x. It can be differentiated by typing **diff(f(x),x);** (here, you must type $f(x)$ and not just f). The disadvantage in using this syntax is that **diff** returns an expression instead of a function. Thus you must use the more cumbersome **subs** command for evaluating the derivative at particular points.

3.4   Implicit Differentiation

Background Information: Read Section 3.7 in Stewart's **Calculus**.

Up until now, we have discussed functions and expressions that are defined explicitly, which means that the dependent variable, such as y or f, appears on one side of an equation and an expression of x appears on the other side (e.g., $y = x^2$). In this chapter, we consider functions and expressions that are given *implicitly*. This means that the variables are often mixed together in the equation. An example is the Folium of Descartes, which is given by the equation $3xy = x^3 + y^3$. Note that it would be very difficult to solve for the variable y explicitly in terms of x. Nevertheless, plots and derivatives can still be obtained.

To plot this equation, the **implicitplot** command is used.

```
> with(plots): eq:=3*x*y=x^3+y^3;
```
$$eq := 3\,x\,y = x^3 + y^3$$

```
> implicitplot(eq,x=-3..3,y=-3..3, scaling=constrained);
```

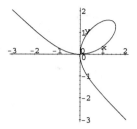

Note that this plot contains a loop, which cannot be described globally as the graph of one function $y = y(x)$. However, near most points the plot is the graph of one function. For example, the lower piece of the loop over the interval $-1 \leq x \leq 1$ is the graph of one function $y(x)$. Finding a formula for $y(x)$ involves solving the equation $3xy = x^3 + y^3$ for y in terms of x. This is difficult since this equation involves a cubic. It is possible to find numerical values of $y(x)$ at specific values of x. For example, the value of y at $x = 1.5$ can be found by using **fsolve**.

> x:=1.5; fsolve(eq,y,y=1..2);

$$x := 1.5$$

$$1.500000000$$

Therefore, $y = 1.5$ at $x = 1.5$. The range **1..2** is inserted into the **fsolve** command so that the larger of the two possible values of y is obtained (see the plot).

A plot over a small range that limits the range of x and y also reveals that the plot satisfies the vertical line test near $x = 1.5$.

> x:='x': implicitplot(eq,x=1.25..1.75,y=1.25..1.75,scaling=constrained);

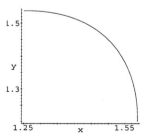

Indeed, over a very small plot range, the graph usually looks like a straight line (the tangent line), and labels on the axes may overprint one another.

> implicitplot(eq,x=1.49..1.51,y=1.49..1.51,scaling=constrained);

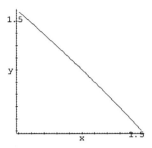

Implicit differentiation is the procedure used to find the derivative of an implicitly defined function or expression by using an equation. The following sequence of commands used for implicit differentiation will be applied to the Folium, but this sequence of commands also applies equally well to other implicitly defined expressions.

First, the variable *eq* is assigned to the equation. Then y is replaced by $y(x)$ (the unknown function).

> **eq:=3*x*y=x^3+y^3; subs(y=y(x),%);**
$$eq := 3\,x\,y = x^3 + y^3$$

$$3\,x\,y(\,x\,) = x^3 + y(\,x\,)^3$$

Next, take the derivative of both sides of the equation. Each side of this equation is an expression in x, so the **diff** command must be used to differentiate it.

> **diff(%,x); solve(%,diff(y(x),x));**
$$3\,y(\,x\,) + 3\,x\,\left(\frac{\partial}{\partial x}\,y(\,x\,)\right) = 3\,x^2 + 3\,y(\,x\,)^2\,\left(\frac{\partial}{\partial x}\,y(\,x\,)\right)$$

$$-\frac{3\,y(\,x\,) - 3\,x^2}{3\,x - 3\,y(\,x\,)^2}$$

The symbol $\dfrac{\partial}{\partial x}$ stands for derivative with respect to x. In the first step, Maple uses the chain rule to take the derivative. Then, in the second step, it isolates y' with the **solve** command.

Since the derivative has now been taken, there is no further need to use the notation $y(x)$ to emphasize the fact that y depends on x. Therefore, y is substituted for $y(x)$ and then the derivative is simplified (and assigned to the variable Dy).

```
> subs(y(x)=y,%); Dy:=simplify(%);
```

$$-\frac{3\,y - 3\,x^2}{3\,x - 3\,y^2}$$

$$Dy := \frac{-y + x^2}{x - y^2}$$

The expression Dy is the *implicit derivative*, and its numerical value can be determined at any point (x, y) by inserting the x- and y-values into its formula. For instance, the slope at the point $(3/2, 3/2)$ is

```
> m:=subs({x=3/2,y=3/2},Dy);
```

$$m := -1$$

This result agrees with the preceding graph.

3.5 Linear Approximation

Background Information: Read Sections 2.6 and 3.10 in Stewart's **Calculus**.

The equation of the tangent line of the previous curve can be determined from the information that its slope is -1 and that it passes through the point $(3/2, 3/2)$.

```
> a:=3/2: b:=3/2: m:=-1:
> tang:=m*(x-a)+b;
```

$$tang := -x + 3$$

The tangent line represents the linear approximation to the graph of the original function. The above expression for the tangent line can be converted to a function (called L).

```
> L:=unapply(tang,x);
```

$$L := x \rightarrow -x + 3$$

The linear function L approximates the values of y for x near 1.5. To plot both the curve and its tangent line, use the **display** command described in Chapter 2.

```
> p1:=plot(L(x),x=1..2):
> eq:=3*x*y=x^3+y^3:
> p2:=implicitplot(eq,x=1..2,y=1..2,scaling=constrained):
> with(plots):
> display([p1,p2]);
```

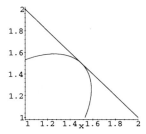

Let's compare the values of y and $L(x)$ at $x = 1.51$. To find the value of y at $x = 1.51$, use **subs** and **fsolve**.

```
> subs(x=1.51,eq); fsolve(%,y,y=1.4..1.6);
```

$$4.53\,y = 3.442951 + y^3$$

$$1.489436032$$

On the other hand, $L(1.51) = 1.49$, as you can easily check. This value is very close to the actual value of $y = 1.489436032$ given above.

3.6   Summary

Differentiation

- Use a Maple function to compute a difference quotient.

- Compute the slope of a tangent line by the **Limit–value** syntax applied to the difference quotient.

- Write a Maple expression for the line tangent to the graph of $y = f(x)$. Display both the curve and the line tangent to it on the same graph.

- The **Limit–value** syntax can be used to calculate the derivative as a limit of difference quotients.

- Given a Maple function f, use **D(f)** to compute the derivative function. Know how to evaluate it.

- Given a Maple function f, use **D@@n** to compute higher derivative functions. Know how to evaluate them.

- Given a Maple expression **expr**, use **diff(expr,x)** to calculate an expression for the derivative. Use **subs** to evaluate it.

- Given a Maple expression **expr**, use **diff(expr,x$n)** to calculate the nth order derivative of **expr with respect to** x. Use **subs** to evaluate it.

Implicit Differentiation

- Use **diff(eq,x)** to take the derivative of a Maple equation.

- Maple expressions and Maple functions both correspond to the idea of an explicit mathematical function, $y = f(x)$, where there is an independent variable x and a dependent variable y. A vertical line crosses the graph at most once for a given x, and there is an explicit rule to compute y given x.

- Use **implicitplot** in the **with(plots)** library to plot a Maple equation.

- At points where a vertical line crosses the graph more than once, we cannot represent the graph with a single function.

- Regardless of whether Maple can explicitly solve an equation to isolate y, if the vertical line test works in a box, we say that y is defined implicitly as a mathematical function of x, i.e., $y = y(x)$ for some possibly unknown formula $y(x)$.

- The implicit derivative is the derivative of that (possibly unknown) formula. Maple can find this derivative, even if it cannot solve for y.

- Compute the implicit derivative. (Comments are inserted in Maple commands by means of the # symbol. They are not needed for the command to work, but they remind the user of the logic behind a Maple command.)

```
> subs(y=y(x),eq);     # View y as an implicit function of x.
> diff(%,x);     #Differentiate both sides by x.
> solve(%,diff(y(x),x));     # Solve for Dy.
> subs(y(x)=y,%);     # Suppress the dependence on x.
> Dy:=simplify(%);     # Simplify the derivative.
```

Linear Approximation

- The linear approximation to a curve at a point (x_0, y_0) on the curve is the function given by the formula for the tangent line.

```
> m:=subs({x=x0,y=y0},Dy);   # Insert point in derivative.
> tang:=m*(x-x0)+y0;   # Point-slope formula for tangent line.
> L:=unapply(tang,x);   # Convert expression to function.
```

3.7   Exercises

1. Differentiate these expressions in Maple.

    (a) $\dfrac{t^2 + t}{t^3 - 1}$

    (b) $\cos^2 (t^3 + 1)$

2. Assign the the expression $r^3 + \sin (r) \cos (r)$ to the variable f. Then differentiate f with respect to r.

3. Assign the expression $a \sin (x^2) + b \cos (x^2)$ to the variable f. Differentiate f with respect to x (here, a and b are constants). Now differentiate f with respect to a (with x and b constant).

4. For each of Exercises 1, 2, and 3, define the given expression as a function and then use the **D** syntax to take its derivative.

5. Define the functions $f(x) = x^3 + 3x + 7$ and $g(x) = \tan (x^2 + 1)$ in Maple. Then use Maple to differentiate $f(x)g(x)$, $f(x)/g(x)$, $(f \circ g)(x)$, and $(g \circ f)(x)$. Here, the composition $f \circ g$ can be entered in Maple as **f@g**.

6. Find the equation of the tangent line to the function $f := x \to x^4 - x^3$ at the point $x = 1.2$. Plot both the graph of the function and the graph of the tangent line on the same coordinate axes.

7. Find the angle of inclination of the tangent line in Exercise 6. *Hint:* Recall that the tangent of the angle of inclination is the slope of the line. So this problem amounts to taking the inverse tangent of the slope of the line in Exercise 6. In Maple, the syntax for the inverse tangent is **arctan( );**

8. Find the point of intersection of the line $y = 2x + 1$ and the curve $y = \sqrt{24.5 - x^2}$. Then find the angle between the tangent lines at this point of intersection.

9. *This exercise does not involve Maple. It is a warm-up for the next exercise, which does involve Maple.* Suppose the curve pictured here represents a plot of distance versus time for an object (time is the horizontal axis and distance is the vertical axis). From this plot, determine the approximate time T when the instantaneous velocity of the object equals the average velocity of the object over the interval $0 \le t \le T$.

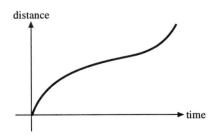

10. Consider the following data, which represent the position (in meters) of an object at various times (in seconds).

Time	0.10	0.20	0.30	0.40	0.50	0.60	0.70	0.80	0.90
Position	0.25	0.40	0.65	0.84	0.99	1.10	1.20	1.38	1.58

Plot this data set using Maple. *Hint:* Enter the data as a list, with alternating times and positions, i.e., **data:=[[0.1,0.25],[0.2,0.4], ...];** as done in Chapter 2. Then type **plot(data);**. This will connect all the data points with line segments. To obtain a plot of data points without the connecting line segments, enter the command **plot(data,style=point);**. From this plot, estimate the time T at which the instantaneous velocity is equal to the average velocity over the time interval $0 \leq t \leq T$.

11. *Newton's Method.* We have already seen examples of solving equations, such as $x^3 + x - 1 = 0$, numerically with the **fsolve** command. The point of this exercise is to explore one algorithm, called Newton's method, which is often used to solve equations numerically. The basic idea behind Newton's method is as follows (for more details, see the section on Newton's method in your text). To solve the equation $f(x) = 0$, pick a starting point, x_0, near the solution to $f(x) = 0$ (for example, from the graph, choose the closest integer to the solution). Generally, the x-intercept of the tangent line to $y = f(x)$ at x_0 is closer to the solution than is x_0. We denote this x-intercept by x_1.

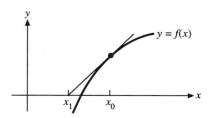

After computing the equation of the tangent line and finding its x-intercept, we obtain

$$x_1 = x_0 - \frac{f(x_0)}{f'(x_0)}.$$

Then we iterate this process. This leads to a sequence of points

$$x_{n+1} = x_n - \frac{f(x_n)}{f'(x_n)} \quad \text{for} \quad n \geq 0$$

which approaches the solution to the equation $f(x) = 0$ as n gets bigger (in fact, n usually does not have to get too large).

(a) This algorithm is programmed into Maple with the following example. Define the function $f := x \to x^3 + x - 1$ (enter this function into Maple). A graph of this function reveals that a root to the equation $f(x) = 0$ exists in the interval $0 \le x \le 1$.

Define the following function, called *Newton*, which yields the next iteration of Newton's algorithm.

```
> Newton:=x->evalf(x-f(x)/D(f)(x));
```

Using $x = 0$ as a starting point, **Newton(0);** returns the first iteration of Newton's method. Now repeatedly execute the command **Newton(%);** until the output no longer changes. (Use copy and paste to obtain multiple copies of the command.) Compare your answers with the result of Maple's **fsolve** command.

(b) The following set of commands executes five iterations of Newton's method with $x_0 = 0$.

```
> x:=0;
> for n from 1 to 5
> do
> x:=evalf(x-f(x)/D(f)(x));
> od;
```

The commands between the **do** and the **od;** will be executed 5 times (with the variable n as the counter). The value of x after this program is executed will contain an approximation to the solution of the equation $f(x) = 0$. Compare this approximation to the solution obtained by using Maple's **fsolve** command.

(c) A variation of the above problem is to execute Newton's method until the difference between the most recent and next most recent values in the sequence of approximations (i.e., $|x_{n+1} - x_n|$) is less than some pre-assigned small number, such as 10^{-7}. One way to do this is to keep track of the most recent and next most recent values in the sequence of approximations by the variables *xnew* and *xold* and then assign the variable *tol* (for tolerance) to their difference $|xnew - xold|$. The algorithm should continue to execute until *tol* is less than the pre-assigned value. To implement this in Maple, first assign the value 1 to the variable *tol* and then use the following modification of the **for** statement in the above code.

```
> for n from 1 to 20 while (tol>=10^(-7))
> do
> ...
> od;
```

This syntax will execute the statements between the **do** and the **od** as long as the variable *tol* exceeds 10^{-7} or until the counter n reaches 20. (The choice of 20 is somewhat arbitrary. Its purpose is to keep the program from running indefinitely if *tol* never gets below 10^{-7}.) For this exercise, construct appropriate Maple commands that belong between the **do** and the **od** statements in the above to implement Newton's method on the equation $x^3 + x - 2 = 0$ until $|xnew - xold|$ is less than 10^{-7}.

12. A cylindrical can with a top and bottom is to contain 1000 cubic centimeters. Find the dimensions of the can if its surface area is 600 square centimeters. *Note:* There are two answers.

13. The bottom of a trough is constructed from a 2 foot by 10 foot rectangular piece of metal by bending it so that the 2 foot width forms an arc of a circle (see the figure). If the volume of the trough is 4 cubic feet, find the angle t subtended by the arc. Use several iterations of Newton's method to solve the resulting equation for the volume (see the hint below) and compare your answer to the one obtained by using **fsolve**.

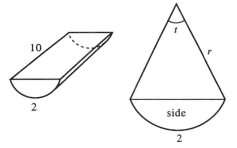

Hint: Suppose that r is the radius of the circular arc subtended by the angle t. Note that $rt = 2$ and that the cross sectional area of the trough is

$$\tfrac{1}{2}r^2 t - r^2 \sin(t/2) \cos(t/2)$$

(Think of the area of the cross section as the area of the circular arc minus the area of a triangle.) Use these equations to derive the equation of the volume of the trough

$$volume = \frac{20(t - \sin(t))}{t^2}$$

14. A pulley consists of an 18 inch band tautly wrapped around two wheels of radius 1 inch and 2 inches, respectively, as shown in the diagram. Find the length of the straight pieces of the band that are not in contact with

either wheel.

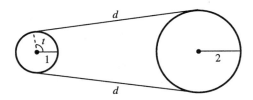

Hint: Let t be the angle of the sector of the smaller wheel (measured from the horizontal) that does *not* make contact with the band. The angle t also represents the angle of the sector of the larger wheel that *does* come in contact with the band (why?). Using the fact that the perimeter of the band is 18 inches, derive the equation

$$2t + 2\pi + 2d = 18$$

Now derive the equation $\tan(\pi - t) = d$ (use trigonometry and similar triangles). Now solve this equation for d and substitute it into the first equation. Solve the first equation using **fsolve**. Use several iterations of Newton's method to solve this equation and compare your answer to the one obtained by using **fsolve**.

15. By following the steps given below, show that the derivative obtained implicitly (without solving for y) is the same as the derivative obtained explicitly by solving for y.

    (a) Consider the equation $x^2 + y^2 = 1$. Use the procedure described in the text to compute the implicit derivative and label it Dy.

    (b) Use Maple to solve the equation for y and label the result as *sol*.

    (c) Use the maple command **subs(y=sol[1],Dy);** to replace y by its explicit version.

    (d) Use the Maple command **diff(sol[1],x);** to take the derivative explicitly.

    (e) Compare your answers in (c) and (d).

    (f) Repeat your work with **sol[2]**.

16. One of the virtues of the implicit derivative process is that, given an equation relating x and y, it is not necessary to explicitly solve for y in order to compute y'. On the other hand, the answer is sometimes not as nice as an explicit answer.

    (a) Consider the equation $eq := y^2 - 3xy + 2x^2 = 0$. Use the procedure described in the text to compute the implicit derivative.

    (b) Now use **with(plots):** and **implicitplot** to have Maple draw the graph. Give a better answer for the derivative than the one you got in (a).

(c) Try **factor(eq);**. Now explain why the answer in (b) is true.

(d) Without executing a series of Maple commands, use the insight that you have gained from Maple graphs and **factor** to give the linear approximation $L(x)$ at any point (a, b) on the graph, except the origin. Can you find the linear approximation at the origin?

17. In the case of the Folium of Descartes, $3xy = x^3 + y^3$, for *most* points (x, y) on the graph there is a plot range for which the graph *locally* passes the vertical line test. However, there are two exceptions where there are vertical tangents.

   (a) Using the **implicitplot** command, click with your mouse to find approximate floating point decimal coordinates for the two points on the graph that do not have this property.

   (b) Refine your guess as to the coordinates of these two points by noting that, to solve for y', you have to divide by $x - y^2$. This division could fail if $x - y^2 = 0$ for a point (x, y) on the curve. Use **fsolve({x-y^2=0,eq},{x,y});**. To find *both* points you'll need to use the guess obtained from clicking on the plot and limit the range of the x- and y-searches, since an unrestricted **fsolve** finds only one point.

   (c) Use **solve({x-y^2=0,eq},{x,y});** to find the *exact* x- and y-coordinates of the two points. (The answer is phrased in terms of **RootOf**, but you should be able to figure out what that means once you see the output in this case.) Confirm that your exact value is correct by using the **evalf** command on the exact answer to see if it matches the **fsolve** answer.

18. For each of the following, use **implicitplot** to plot the given equation. Then find the y-values of the lower piece of its plot at $x = 1, 1.25, 1.5, 1.75, 2$. Compute the slope of the tangent line at $x = 1$ by implicit differentiation and compare this value to the slope of the chord between $x = 1$ and $x = 1.25$.

   (a) $(x^2 + y^2)^2 = 16(x^2 - y^2)$

   (b) $x^2 y + xy^2 = 16$

19. Graph the equations $y = x^2$ and $x^4 + 4x + 2y^2 + 6y = 12$. Find the point(s) of intersection of these graphs. For each point of intersection, find the acute angle between the tangent lines of both equations.

Chapter 4

Applications of Differentiation

Background Information: Read Section 3.9 and Chapter 4 in Stewart's **Calculus**.

A number of Maple commands are used in this chapter to solve some classic problems associated with differentiation; namely, related rate problems, graphical analysis problems and max/min problems .

4.1 Related Rates

Background Information: Read Section 3.9 in Stewart's **Calculus**.

In many situations, two or more quantities are related by some formula and are changing with time. In a related rates problem, the goal is to determine how fast one of the quantities is changing when the rates of change of the other quantities are known.

The procedure for solving this type of problem can be outlined as follows:

1. Determine the two quantities that are changing with time.

2. Find an equation that relates the two quantities (this may involve known formulas from physics or geometry).

3. Enter the equation into Maple, replacing the two quantities, say u and v, by $u(t)$ and $v(t)$ to reflect the fact that the quantities change with time.

4. Take the implicit derivative of the equation with respect to time t.

5. Solve the resulting equation for the desired derivative.

6. Find numeric values for each of the quantities in the resulting solution, possibly by using the equation in step 3, and substitute them into the result.

The following is a typical example.

Example. For a concave lens with focal length 30 cm, the formula from optics $\frac{1}{30} = \frac{1}{u} + \frac{1}{v}$ expresses the relationship between the distance u of an object from the lens and the distance v of its image from the lens. Suppose an object is moving towards the lens at a rate of 3 cm/sec. Find the rate at which the image is receding from the lens when the object is 90 cm away.

Solution. The two quantities that change with time are the two distances, u and v. The equation that relates them is the optics formula given in the problem. This equation is entered into Maple with $u(t)$ and $v(t)$, signifying that the distances are time-dependent.

```
> 1/30=1/u(t)+1/v(t);
```
$$\frac{1}{30} = \frac{1}{u(t)} + \frac{1}{v(t)}$$

This equation is differentiated with respect to t.

```
> diff(%,t);
```
$$0 = -\frac{\frac{\partial}{\partial t}u(t)}{u(t)^2} - \frac{\frac{\partial}{\partial t}v(t)}{v(t)^2}$$

The resulting equation is solved for $v'(t)$.

```
> vrate:=solve(%,diff(v(t),t));
```
$$vrate := -\frac{\left(\frac{\partial}{\partial t}u(t)\right)v(t)^2}{u(t)^2}$$

The numerical value of v is found.

```
> 1/30=1/90+1/V; imagelength:=solve(%,V);
```
$$\frac{1}{30} = \frac{1}{90} + \frac{1}{V}$$

$$imagelength := 45$$

The answer is found by substituting the values of $u'(t) = -3$, $u(t) = 90$, and $v(t) = imagelength$ into the variable *vrate*.

```
> imagerate:=subs({diff(u(t),t)=3,v(t)=imagelength,u(t)=90},vrate);
```
$$imagerate := \frac{3}{4}$$

4.2 Local Max/Min

Background Information: Read Section 4.1 in Stewart's **Calculus**.

In this section, you will learn how to use the **plot**, **diff**, and **fsolve** commands to find local maxima and minima of differentiable functions.

Recall that if f is a differentiable function on an open interval, then its derivative must vanish (i.e., $f' = 0$) at each local maximum or minimum. So the strategy for finding local maxima and minima is to first plot the function to get an approximate idea of the location of the maxima and minima. Then use the **solve** or **fsolve** command to find the solutions of the equation $f' = 0$.

Example. Plot the expression $f = x^3 + 0.2x^2 - x$ and find the location of the local maxima and minima.

Solution. Define f to be the expression $x^3 + 0.2x^2 - x$.

> f:=x^3+0.2*x^2-x;
$$f := x^3 + .2\,x^2 - x$$

Set Df equal to its derivative.

> Df:=diff(f,x);
$$Df := 3\,x^2 + .4\,x - 1$$

Use **plot(f,x=-2..2);** to observe that the graph of f has a local max between -1 and 0 and a local min between 0 and 1.

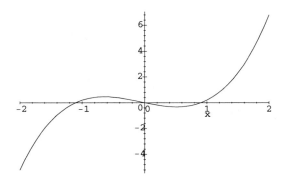

To find the precise location of the local min and local max, the equation $Df = 0$ must be solved for x. The local max that is between -1 and 0 is labeled *xmax*

> **xmax:=fsolve(Df=0,x,x=-1..0);**

$$xmax := -.6478531925$$

and the local min that is between 0 and 1 is labeled *xmin*.

> **xmin:=fsolve(Df=0,x,x=0..1);**

$$xmin := .5145198591$$

Inserting these values back into the original expression f will give the corresponding y-coordinates of the local min and local max.

> **subs(x=xmax,f); subs(x=xmin,f);**

$$.4598830456$$

$$-.3253645270$$

Therefore, the local max is the point $(-.6478531925, .4598830456)$ and the local min is $(.5145198591, -.3253645270)$.

Note. If f is defined as a function $(f:= x \rightarrow x^3 + 0.2x + x^2)$, then the x-coordinate of the local max can be found as follows.

> **xmax:=fsolve(D(f)(x)=0,x,x=-1..0);**

$$xmax := -.6478531925$$

The corresponding y-coordinate can then be obtained by typing

> **f(xmax);**

$$.4598830456$$

Of course, the above example could have been done by hand computation. However, the example in the next section, and many of the exercises are too complicated for hand computation.

4.3   Graphical Analysis

Background Information: Read Sections 4.3 through 4.6 in Stewart's **Calculus**.

In this section, we consider the graph of the expression

$$f := \frac{e^x}{x^3 + x - 1 + 0.2e^x}$$

in detail; enter this expression into Maple with e^x entered as **exp(x)**. The goal is to produce an accurate graph, locating horizontal and vertical asymptotes, local max/min, and inflection points.

First, plot the graph over the interval $-5 \le x \le 15$.

```
> f:=exp(x)/(x^3+x-1+0.2*exp(x)); plot(f,x=-5..15,y=-10..10);
```

$$f := \frac{e^x}{x^3 + x - 1 + .2 e^x}$$

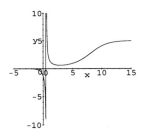

The y-range is restricted to $-10 \le y \le 10$ to obtain a reasonable plot. From the graph, it appears that there is a vertical asymptote between $x = 0$ and $x = 1$. Its location can be pinpointed by locating the root of the denominator of f. In Maple, the denominator of an expression f can be referred to by the command **denom(f)**, which saves typing in cases where the denominator is complicated.

```
> fsolve(denom(f)=0,x=0..1);
```

$$.5213890506$$

The lines $y = 0$ and $y = 5$ appear to be horizontal asymptotes as $x \to -\infty$ and $x \to \infty$, respectively. This can be verified by taking limits (note that when x is a large negative number, $f \approx 0$ and, when x is a large positive number, $f \approx \frac{e^x}{0.2e^x} = 5$).

```
> Limit(f,x=-infinity); value(%);
```

$$\lim_{x \to (-\infty)} \frac{e^x}{x^3 + x - 1 + .2 e^x}$$

$$0$$

> **Limit(f,x=infinity); value(%);**

$$\lim_{x \to \infty} \frac{e^x}{x^3 + x - 1 + .2\,e^x}$$

$$5.$$

From the plot, there appears to be a local minimum between $x = 2$ and $x = 4$. Its location can be pinpointed by taking the derivative and locating its root.

> **Df:=diff(f,x): a:=fsolve(Df=0,x,x=2..4);**

$$a := 2.893289196$$

The value of the expression f at the local minimum is 0.6073428968, which can be determined by using the command **evalf(subs(x=a,f));**.

It is instructive to replot both f and its derivative on the same coordinate axes. We focus on the part of the graph of f over the interval $0.5 \le x \le 15$, since the local minimum of f is located there.

> **plot({f,Df},x=0.5..15,y=-10..10);**

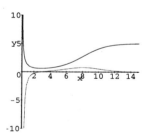

Note that f decreases on the interval $0.5 < x < 2.89$ and the derivative of f is negative on this same interval. Likewise, f increases on $x > 2.89$ and the derivative of f is positive on this same interval. The graph of f flattens as x gets large. This corresponds to the observation that the graph of the derivative approaches $y = 0$ when x gets large.

There also appears to be an inflection point (a point where the concavity of the graph switches) between $x = 6$ and $x = 10$. Its location can be pinpointed by taking the second derivative (the derivative of the first derivative, Df) and finding its root.

> **DDf:=diff(Df,x): b:=fsolve(DDf=0,x,x=6..10);**

$$b := 8.131398912$$

The value of the expression f at $x = b$ is 2.775853638, which can be found by using the command **evalf(subs(x=b,f));** So the point of inflection is approximately $(8.131, 2.776)$.

We now plot the expression f and its second derivative on the same coordinate axes.

> **plot({f,DDf},x=0.5..15,y=-1..5);**

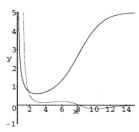

The scale for the y-coordinate is changed to more clearly display the graphs. Note that f is concave up and the second derivative is positive on the interval $0.53 < x < 8.13$. The graph of f is concave down and the second derivative is negative on the interval $x > 8.13$. Also note that the graph of f is steepest where the first derivative is largest (see previous graph). This point, $x = 8.13$, is where the second derivative is zero.

4.4   Designer Polynomials

Background Information: Read Section 4.3 and 4.5 in Stewart's **Calculus**.

Here is an example that combines the use of many of the Maple commands introduced up to this point. The problem is to find the coefficients of the cubic polynomial

$$f(x) = ax^3 + bx^2 + cx + d$$

that has a relative minimum at $(-2, 0)$ and a relative maximum at $(3, 4)$.

Solution. Four equations are needed for the four unknowns a, b, c and d. These equations will be labeled *eq1* through *eq4* for later reference. After defining f as a Maple function, the first two equations can be obtained from the information $f(-2) = 0$ and $f(3) = 4$.

```
> f:=x->a*x^3+b*x^2+c*x+d;
> eq1:=f(-2)=0;
> eq2:=f(3)=4;
```

$$f := x \to a\,x^3 + b\,x^2 + c\,x + d$$

$$eq1 := -8\,a + 4\,b - 2\,c + d = 0$$

$$eq2 := 27\,a + 9\,b + 3\,c + d = 4$$

The remaining two equations can be obtained from the fact that the derivative of f must vanish at $x = -2$ and $x = 3$ (because f has a local min and max at these points).

```
> eq3:=D(f)(-2)=0; eq4:=D(f)(3)=0;
```

$$eq3 := 12\,a - 4\,b + c = 0$$

$$eq4 := 27\,a + 6\,b + c = 0$$

Now these four equations are solved for a, b, c, and d.

```
> sol:=solve({eq1,eq2,eq3,eq4},{a,b,c,d});
```

$$sol := \left\{ a = \frac{-8}{125}, d = \frac{176}{125}, b = \frac{12}{125}, c = \frac{144}{125} \right\}$$

(Note: Your solutions may appear in a different order.) The label *sol* (short for solution) is used for further reference. Note the use of the curly braces { } to enclose *eq1, eq2, eq3, eq4* and *a, b, c, d*. Maple requires the use of curly braces when enclosing a set of objects. The output *sol* is also in the form of a set. The first element in this set can be referred to by typing **sol[1]** and so forth. To substitute values of *a, b, c, d* back into the original function, type

```
> g:=subs(sol,f(x));
```

$$g := -\frac{8}{125}\,x^3 + \frac{12}{125}\,x^2 + \frac{144}{125}\,x + \frac{176}{125}$$

Plot the graph of g to see if it has a local min at $(-2,0)$ and a local max at $(3,4)$ (do this!).

4.5 Finding Absolute Extremes of a Function on an Interval

Background Information: Read Sections 4.1 and 4.7 in Stewart's **Calculus**.

In this section, Maple is used to help find the absolute maximum and/or minimum of an expression or a function on an interval. The method compares the values of the function or expression at its critical points on the given interval with those at the endpoints of the interval. This is illustrated by the following example.

Example. Find the absolute maximum and minimum of the expression $f = \ln(x) - 4x^2 + x^3$ on the interval $1 \le x \le 3$.

Solution. First, enter f as an expression in Maple and then plot it over the interval $1 \le x \le 3$.

> f:=ln(x)-4*x^2+x^3; plot(f,x=1..3);
$$f := \ln(x) - 4x^2 + x^3$$

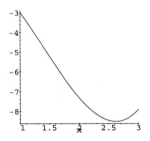

From the plot, it appears that the minimum of f is located at a critical point near $x = 2.5$ and that the maximum of f is located at the left hand endpoint $x = 1$. To find the critical point, set the derivative of f equal to zero and solve.

> Df:=diff(f,x); xmin:=fsolve(Df=0,x,x=2..3);
$$Df := \frac{1}{x} - 8x + 3x^2$$

$$xmin := 2.618033989$$

The values for f at the endpoints $x = 1$ and $x = 3$ as well as at $x = xmin$ are approximately -3, -7.9014, and -8.5097, respectively (determined by using the **subs** and **evalf** commands). So, the maximum of f is -3 at $x = 1$ and the minimum of f is -8.5097 at $x = 2.618$.

If the interval is open (does not contain its endpoints) or if the interval is infinite in length, then there may not be an absolute maximum or minimum. As an example, consider a plot of the expression $f = \dfrac{x^4 + 2x + 3}{x^2}$ over the interval $x > 0$.

> f:=(x^4+2*x+3)/x^2; plot(f,x=0..20,y=-100..100);
$$f := \frac{x^4 + 2x + 3}{x^2}$$

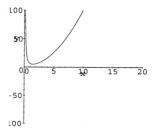

The plot reveals an absolute minimum at a critical point near $x = 1.5$. However, this expression has no absolute maximum, since f approaches infinity as $x \to 0^+$ or $x \to \infty$. The minimum can be found by computing the derivative, **Df:=diff(f,x);** and then setting the derivative equal to zero, **fsolve(Df=0,x,x=1..2);**. The minimum of f is approximately 4.9087, occurring at $x = 1.4526$.

4.6   The Most Economical Tin Can

Background Information: Read Section 4.7 in Stewart's **Calculus**. Related material can be found in the Applied Project: *The Shape of a Can.*

In this section, an example of an applied max/min problem that cannot be easily solved using hand computation is presented. The discussion will focus on the problem solving strategy and the role of Maple in that strategy rather than on the Maple syntax itself.

For an applied max/min problem, you must first translate the problem into mathematics (Maple will not help you here) and then solve the resulting calculus problem (this is where Maple can help).

For example, consider a cylindrical tin can which is to be constructed by joining the ends of a rectangular piece of material to form the cylindrical side and then attaching circular pieces to form the top and bottom. There are seams around the perimeter of the top and bottom and there is one seam down the side surface (where the ends of the rectangle join together). Suppose the volume of the can is 1000 cubic centimeters. Also suppose that the cost of the material is $1.00 per square meter and the cost of the seam is $0.20 per meter. Find the dimensions of the can that will minimize its cost.

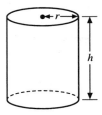

Solution. Let r be the base radius of the can and h be the height of the can. The surface area of the can is

> s:=2*Pi*r^2+2*Pi*r*h;
$$s := 2\pi r^2 + 2\pi r h$$

The first term on the right represents the area of the top and bottom. The second term represents the area of the side (which has the same area as a rectangle of length $2\pi r$ and height h).

The total length of the seam is

> l:=4*Pi*r+h;
$$l := 4\pi r + h$$

We will use centimeters as our units. The cost of the material is $a = 0.01$ cents per square centimeter (100 cents divided by the 10000 square centimeters in a square meter). The seaming cost is $b = 0.2$ cents per centimeter. The cost of the material is as and the cost of the seams is bl. So the total cost of the can is

> c:=a*s+b*l; a:=0.01; b:=0.2;
$$c := a\left(2\pi r^2 + 2\pi r h\right) + b\left(4\pi r + h\right)$$

$$a := .01$$

$$b := .2$$

This cost expression has both r and h as variables. Eliminate the variable h by using the fact that the volume of the box is $1000 = \pi r^2 h$ (volume = base × height). Entering this fact into Maple yields a cost expression in terms of one variable only, namely r.

```
> h:=1000/(Pi*r^2); c;
```

$$h := 1000\,\frac{1}{\pi\,r^2}$$

$$.02\,\pi\,r^2 + 20.00\,\frac{1}{r} + .8\,\pi\,r + 200.0\,\frac{1}{\pi\,r^2}$$

This cost expression is to be minimized over the interval $r > 0$. Plotting it, we observe that a minimum occurs somewhere between $r = 3$ and $r = 6$.

```
> plot(c,r=1..10);
```

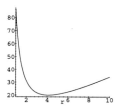

To minimize the cost expression, set its derivative Dc equal to zero and solve for r, labeling the solution as *rbest*.

```
> Dc:=diff(c,r): rbest:=fsolve(Dc=0,r,r=3..6);
```

$$rbest := 4.104217144$$

The answer is the best value for the radius (in centimeters). Substituting this value into the expression for h yields

```
> evalf(subs(r=rbest,h));
```

$$18.89685223$$

4.7   Summary

Related Rates

- Related rates problem: suppose two quantities related by a Maple equation, *eq*, are both changing with time. Given the rate at which one changes, find the rate at which the other changes.

- Procedure: read the problem; identify quantities that change; produce an equation that relates the two quantities, possibly using known formulas from geometry, physics, or the like.

- Example: suppose the quantities are denoted by u and v, and the equation relating them, eq, has been entered into Maple. The known rate is u'_0, and u_0 is the value of u.

```
> subs({u=u(t),v=v(t)},eq);    # Quantities u and v are implicit
> functions of time.
> diff(%,t);    # Take the implicit derivative with respect to time.
> vrate:=solve(%,diff(v(t),t));    # Solve for the second rate.
> vsol:=solve(subs(u=u0,eq),v);    # Substitute for u, and solve for v.
> Dv:= subs({diff(u(t),t)=uprime0,u(t)=u0,v(t)=vsol}, vrate);
```

Graphical Analysis of Expressions

Use the following commands for an expression labeled *expr*.

- x-intercept:

```
> plot(expr,x);
                # Plot expr, click on x-intercepts.
> fsolve(expr=0,x,x=a..b);
                # Restrict search, approximate x-intercept.
```

- Local maxima/minima:

```
>plot(expr,x);
                # Plot expr, click on maxima and minima.
>Dy:=diff(expr,x);
                # Compute derivative expression.
>fsolve(Dy=0,x,x=c..d);
                # Restrict search, approximate x-coordinate.
>subs(%,expr);
                # Find y-coordinate of critical point.
```

- Inflection point:

```
>plot(expr,x);
                # Plot expr, click on inflection point.
```

```
>plot(expr,x);>DDy:=diff(Dy,x);
                # Compute second derivative expression.
>fsolve(DDy=0,x,x=e..f);
                # Restrict search, approximate x-coordinate.
>subs(%,expr);
                # Find y-coordinate of inflection point.
```

- Vertical asymptote:

```
>plot(expr,x);
                # Plot expr, click on vertical asymptote.
>simplify(expr);
                # Make expr into single fraction.
>denom(%);
                # Isolate denominator.
>fsolve(%=0,x,x=g..h);
                # Restrict search, approximate zero of

                denominator.
```

- Horizontal asymptote:

```
>Limit(expr,x=infinity); value(%);
                # Look for finite limit on far right.
>Limit(expr,x=-infinity); value(%);
                # Look for finite limit on far left.
```

Graphical Analysis of Functions

Use the following commands for a function labeled *fn*.

- *x*-intercept:

```
>plot(fn);
                    # Plot fn and click on x-intercepts.
>fsolve(fn(x)=0,x,x=a..b);
                    # Restrict search, approximate x-intercept.
```

- Local maxima/minima:

```
>plot(fn);
                # Plot fn, click on maxima and minima.
```

```
>plot(fn);>Df:=D(fn)(x);
                # Compute evaluated derivative function.
>fsolve(Df=0,x,x=c..d);
                # Restrict search, approximate x-coordinate.
>fn(%);
                # Find y-coordinate of critical point.
```

- Inflection point:

```
>plot(fn);
                # Plot fn, click on inflection point.
>DDf:=(D@@2)(fn)(x);
                # Compute evaluated second derivative function.
>fsolve(DDf=0,x,x=e..f);
                # Restrict search, approximate x-coordinate.
>fn(%);
                # Find y-coordinate of inflection point.
```

- Vertical asymptote:

```
>plot(fn);
                # Plot fn, click on vertical asymptote.
>simplify(fn(x));
                # Make evaluated function into single fraction.
>denom(%);
                # Isolate denominator.
>fsolve(%=0,x,x=g..h);
                # Approximate zero of denominator.
```

- Horizontal asymptote:

```
>Limit(fn(x),x=infinity); value(%);
                # Look for finite limit on far right.
>Limit(fn(x),x=-infinity); value(%);
                # Look for finite limit on far left.
```

An Overview of the Maple Process Associated with Maximizing and/or Minimizing an Expression on an Interval

- Plot the Maple expression, **expr,** on the interval $a \leq x \leq b$.

```
> plot(expr,x=a..b);    # Plot expr.
```

- Use the plot and the **fsolve** command to find the x-coordinates of the extrema in the interior of this interval. Then evaluate the expression at these values.

```
> Df:=diff(expr,x);     # Compute derivative expression.
> fsolve(Df=0,x,x=c..d);    # Locate critical point.
> subs(x=%,expr);    # Value of expr at critical point.
```

- Compare extrema in the interior of interval with values of the expression at the endpoints a and b to select absolute extrema.

```
> subs(x=a,expr);    # Value of expr at left endpoint.
> subs(x=b,expr);    # Value of expr at right endpoint.
```

4.8   Exercises

1. When a ray of light hits the surface of a lake, the beam is bent. The equation that governs this effect is due to Willebrod Snell (1591–1626). He noted that
$$\frac{\sin(\theta_1)}{\sin(\theta_2)} = 1.33$$
where 1.33 is the index of refraction of water (relative to air) and θ_1 and θ_2 are the angle of incidence and angle of refraction, respectively, measured from a line perpendicular to the surface of the lake. Suppose that as the sun rises, the angle of incidence of the sunlight decreases at $\pi/12$ radians per hour. When $\theta_1 = \pi/3$, how fast does the fish see the sun rise?

2. Plot the expression $f = 0.25x^4 - 0.913x^3 + x^2 - 0.175x - 0.5$. You may need to adjust the plot ranges to include all relevant aspects of the graph. Find the local max and min and the intervals where f is increasing and decreasing.

3. Find the cubic $ax^3 + bx^2 + cx + d$ that has a local max at $(-1, 2)$ and a local min at $(3, -2)$.

4. Find the cubic polynomial that passes through the points $(-1, -2)$, $(1, 3)$, and $(4, 2)$ and whose derivative at $x = 1$ is 1.7.

5. Find the cubic polynomial with a local maximum at $(-2, 4)$ and a local minimum at $(3, -1)$.

6. Plot an informative graph of $y = \dfrac{x^3 - 6x^2 + 11x - 5.3}{x^3 - 3.1x^2 - 3.2x + 4.21}$. Determine all asymptotes and critical points.

7. Graph the function $y = \dfrac{x \ln(x)}{x^2 + 1}$ for $x > 0$. Find all critical points and inflection points. Find intervals where the function is increasing, decreasing, concave up, and concave down.

8. For each of the following expressions, locate the following:

  - horizontal and vertical asymptotes
  - local max/min
  - inflection points.

  Then plot the given expression, along with its derivative and second derivative. Observe that the intervals where the expression is increasing (decreasing) correspond to the intervals where its derivative is positive (negative). Do a similar analysis comparing concavity for the expression and the sign of the second derivative.

  (a) $\dfrac{10x^2 - x + 1}{10x - 1}$

  (b) $\dfrac{3x^5 + 2x}{x^5 - 3x + 1}$

  (c) $\dfrac{2e^x}{e^x - x^3}$

  (d) $x^2 + 5\cos(x)$

  (e) $\ln(x) - 4x^2 + x^3$

9. Each of the following plots represents the graph of the *derivative* of a function. Draw a plausible graph of the corresponding function and locate (approximately) any local max/min and inflection points.

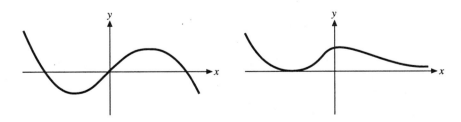

10. Each of the following triplets of plots represents the graph of a function, its derivative, and its second derivative (although not necessarily in that order). Determine the graph that represents the function; the derivative;

the second derivative.

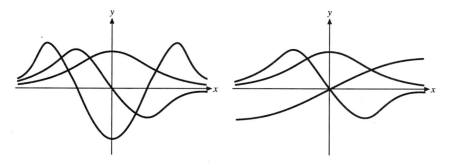

11. (This is a warm-up for the next exercise.) A plot of land has a flat, horizontal southern boundary and an irregular northern boundary. The goal of this problem is to draw a map of the boundary of this plot of land. Since the southern boundary is horizontal, it will be represented by the x-axis with the origin located at its western point. North-south measurements (represented by the variable y) are taken and placed in the following table (each unit represents 100 feet).

x-value	0.0	1.0	2.0	3.0	4.0
y-value	0.0	2.1	3.2	2.4	1.7

The first task is to find the graph of the quadratic polynomial p (a parabola) that contains the first three data points. This is easily accomplished by defining the quadratic polynomial $p(x) = ax^2 + bx + c$ as a function and then by solving the equations $p(0) = 0$, $p(1) = 2.1$, and $p(2) = 3.2$ for the unknowns a, b, and c. The next task is to find the graph of the cubic polynomial q that contains the last three data points and is such that $p'(2) = q'(2)$. This last requirement ensures that the slope of the graph of p at $x = 2$ will agree with that of q at $x = 2$. To do this, enter $q(x) = dx^3 + ex^2 + fx + g$ as a function (if necessary, unassign any previously used labels). Enter the equations $q(2) = 3.2$, $q(3) = 2.4$, $q(4) = 1.7$, and $D(p)(2) = D(q)(2)$; then solve for the unknowns d, e, f, and g.

Graph p and q using the **display** command. As explained in Chapter 2, this is accomplished as follows. Enter the commands **g1:=plot(p(x),x=0..2):** **g2:=plot(q(x),x=2..4):**. Then issue the commands **with(plots): display([g1,g2]);**. This sequence of commands is necessary since p and q are plotted over different intervals.

12. *Smooth off the southern boundary of Texas.* In Exercise 25 in Chapter 2, the boundary of the state of Texas is drawn using line segments. The point of this problem is to use parabolas and cubics to smooth out the part of the southern boundary of the state formed by the Rio Grande River. Enter the relevant Rio Grande data.

```
> rio:=[[0,0],[1,-1.1],[2,-2.5],[3,-2.9],[4,-2.3],[5,-2.8],[6,-4.4],
> [7,-5.8],[8,-6.1]];
```

$$rio := [[0,0],[1,-1.1],[2,-2.5],[3,-2.9],[4,-2.3],[5,-2.8],$$
$$[6,-4.4],[7,-5.8],[8,-6.1]]$$

Here the origin is the western corner of Texas (near El Paso) and the x-axis is the extension of the east-west border between New Mexico and Texas. Each unit represents approximately 69 miles.

To find functions that smooth out the Rio Grande, proceed as in Exercise 11. First, find the parabola $p(x)$ that passes through the three data points $[0,0]$, $[1,-1.1]$, and $[2,-2.5]$. Then find a cubic polynomial q that passes through the next triplet of data points $[2,-2.5]$, $[3,-2.9]$, $[4,-2.3]$ and further satisfies the equation $p'(2) = q'(2)$ (so that the slopes of the graphs of p and q at $x = 2$ are the same). In the same manner, find cubics r and s for the triplets $[4,-2.3]$, $[5,-2.8]$, $[6,-4.4]$ and $[6,-4.4]$, $[7,-5.8]$, $[8,-6.1]$ so that $q'(4) = r'(4)$ and $r'(6) = s'(6)$. Then plot the following on the same coordinate axes: $p(x)$ over $0 \leq x \leq 2$; $q(x)$ over $2 \leq x \leq 4$; $r(x)$ over $4 \leq x \leq 6$; and $s(x)$ over $6 \leq x \leq 8$. Use the **display** command as in Exercise 11.

13. A metal box with a square base and no top holds 1000 cubic centimeters. It is formed by folding up the sides of the flattened pattern pictured here and seaming up the four sides. The material for the box costs \$1.00 per square meter and the cost to seam the sides is 5 cents per meter. Find the dimensions of the box that costs the least to produce.

14. Find the point on the graph of $y = x^2 \tan(x)$ that is closest to the point $(2.1, 0.8)$. *Hint:* Use the distance formula to obtain a function that describes the distance between the point $(2.1, 0.8)$ and a typical point $(x, x^2 \tan(x))$ on the graph of y and then find its minimum using Maple.

15. Find the closest point on the parabola $y = \dfrac{x^2}{4} - 1$ to the point $(2, -1)$.

16. A rectangular movie theater is 100 feet long (from the front screen to the back). The top and bottom of its screen are 40 feet and 15 feet from

the floor, respectively. Find the position in the theater with the largest viewing angle. *Hint:* Arrange a coordinate system with the origin at the floor directly under the screen. You are asked to find the position x where the angle $B - A$ is the largest (see figure). Instead of maximizing $B - A$, it is easier to maximize $\tan(B - A)$; this will lead to the same optimum value of x, since $\tan(\alpha)$ is an increasing function on $-\pi/2 < \alpha < \pi/2$. The subtraction formula for tangent will be needed.

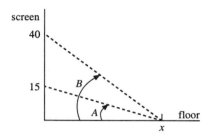

17. Repeat Exercise 16, but this time assume that the floor of the theater has a gentle parabolic slope given by the parabola $y = 0.001x^2$ for $0 \leq x \leq 100$ (the origin is located on the floor directly under the screen).

18. Repeat Exercise 17, but assume that the floor of the theater has a steeper parabolic slope given by the formula $y = 0.004x^2$.

19. A pipeline is to be constructed to connect a station on the shore of a straight section of coast line to a drilling rig that lies 5 kilometers down the coast and 2 kilometers out at sea. Find the minimum cost to construct the pipeline, given that the pipeline costs 4 million dollars per kilometer to lay under water and 2 million dollars per kilometer to lay along shore.

20. Let v_1 be the velocity of light in air and v_2 the velocity of light in water. We know from physics that a ray of light travels from a point A in the air to a point B in the water via a path ACB that minimizes the time taken. The point of this problem is to derive Snell's Law

$$\frac{\sin(\theta_1)}{\sin(\theta_2)} = \frac{v_1}{v_2}$$

where θ_1 (the angle of incidence) and θ_2 (the angle of refraction) are shown in the figure. *Remark:* Use Maple to compute a derivative, if you wish;

the rest of the solution is more easily done by hand.

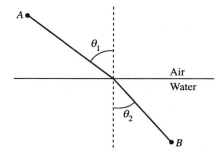

This equation can then be used to calculate θ_2 provided θ_1, v_1, and v_2 are known.

21. Now suppose the medium in Exercise 20 has a curved surface (such as the surface of a glass lens). To be more specific, suppose a beam of light follows the line with the equation $y = -3x + 1$ and strikes a piece of glass whose outside boundary is given by the equation $y = -x^2$. Find the equation of the line that represents the refracted light beam as it travels through the glass. Assume that v_1/v_2 is 1.52.

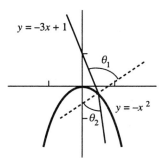

Hint: First, use Maple to find the point of intersection between the light beam and the parabolic surface of the glass. At this point compute the angle between the normal to the parabolic surface and the incoming light beam. This angle is analogous to the angle θ_1 in Exercise 20. Snell's law can now be used to calculate θ_2 and the slope of the refracted light can then be determined.

22. Suppose a glass lens is formed by the intersection of the interiors of the following two ellipses:

$$(x - 4)^2 + y^2/2 = 25$$

and

$$(x + 4)^2 + y^2/2 = 25.$$

A light beam strikes this lens from the right along the horizontal line $y = 2$. Find the equations of the line segments that describe the trajectories of the refracted light beam as it passes through the lens and as it passes through the air on the left side of the lens. Find the location where the refracted light beam hits the negative x-axis.

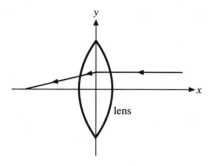

23. Two hallways of width a and b intersect at right angles. What is the length of the longest rigid rod that can be pushed on the floor around the corner of the intersection of these two halls? Carefully explain your steps.

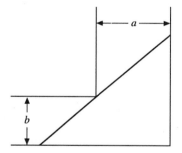

Chapter 5

Integrals

Background Information: Read Chapters 5 and 8 in Stewart's **Calculus**.

In the first section, we use Maple's **leftbox** and **rightbox** commands to visualize the process of approximating the area under a curve. In the second section, we compute some Riemann sums with the help of Maple's **Sum** command. The integration command **Int** is also introduced. The next three sections demonstrate how Maple can be used to perform integration by substitution, integration by parts, and integration by partial fraction decomposition. The final section discusses the midpoint rule, the trapezoidal rule and Simpson's rule for approximating definite integrals.

Since Maple has a built-in integration command, it is reasonable to wonder why we need to discuss techniques of integration. It turns out that Maple cannot integrate everything, but with a little help from us, the number of integrands that Maple is able to integrate in closed form increases. For example, try integrating $f(x) = \sin(x^2) \ln(x)$.

5.1 Visualizing Riemann Sums

Background Information: Read Section 5.1 in Stewart's **Calculus**.

Consider the region under the graph of the function $f := x \to x^2$ over the interval $1 \le x \le 3$ (enter f as a function in Maple). First, divide this interval into n subintervals of equal length. Here, n is typically a large number such as 10 or 100 (later, n will approach ∞). The area under the graph of $f(x) = x^2$ is approximated by the sum of the areas of n rectangles where the base of a typical rectangle is one of the n subintervals and the height is the value of the function $f(x)$ at the left or right endpoint of the subinterval. To get a picture of $n = 10$ rectangles, load the student package and enter the leftbox command as follows:

```
> f:=x->x^2:
> with(student):
> leftbox(f(x),x=1..3,10);
```

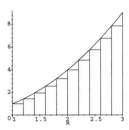

The **leftbox(f(x),x=1..3,10)** command draws the graph of $f(x)$ over the interval $1 \leq x \leq 3$, together with 10 rectangles; the height of each rectangle is the value of the function at the left point of the base of the rectangle. Since the graph of f is increasing over the interval $1 \leq x \leq 3$, the **leftbox** command gives rectangles whose areas under-approximate the area under the graph of f (see the picture above). To obtain rectangles whose heights are determined by the function values at the right endpoints we use the **rightbox** command.

```
> rightbox (f(x),x=1..3,10);
```

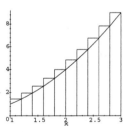

In this case, the **rightbox** command illustrates rectangles that over-approximate the area under the graph of f, since f is an increasing function on the interval $1 \leq x \leq 3$.

The value $n = 10$ can be changed to any positive integer. Try the **leftbox** and **rightbox** commands with $n = 20, 50$, and 100. Note that, as the value of n gets larger, the sum of the areas of the rectangles more closely approximates the area under the graph of f. As n gets larger, the sum of the areas of the **leftbox** rectangles approach the area under the graph from below and the sum of the areas of the **rightbox** rectangles approach the area under the graph from above.

This same rectangle construction can be applied to any interval $a \leq x \leq b$ (not just $1 \leq x \leq 3$). Try different values of a and b (and n). Note that, for negative values of a and b, f is a decreasing function on $a \leq x \leq b$ and therefore **leftbox** rectangles will over-approximate and **rightbox** rectangles will under-approximate the area under the graph of f on such an interval.

For a nonnegative function, such as $f(x) = x^2$, this limit of the sum of the areas of rectangles (as the number of rectangles tends to infinity) is the definition of the integral of f from $x = a$ to $x = b$, and is denoted by the symbol

$$\int_a^b f(x) \, dx$$

5.2   The Computation of the Definite Integral

Background Information: Read Section 5.2 in Stewart's **Calculus.**

We continue with the preceding example of the integral of $f(x) = x^2$ from $x = a$ to $x = b$.

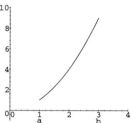

As in the previous section, the interval $a \leq x \leq b$ is divided into n subintervals, each of length

$$\Delta x = \frac{b - a}{n}$$

Let x_i denote the right endpoint of the ith subinterval, where i is a counting index that runs from 1 to n. The area of the ith box is the product of the interval length dx and the height of the box, which is $f(x_i)$. So the sum of the areas of the n-**rightbox** rectangles is

$$\sum_{i=1}^n f(x_i) \, \Delta x$$

Before using Maple to compute this sum, a formula for x_i must be derived. Note the following pattern

$$x_1 = a + \Delta x, \quad x_2 = a + 2 \, \Delta x, \quad x_3 = a + 3 \, \Delta x$$

and so forth. So in general

$$x_i = a + i \, \Delta x$$

Now these formulas can be entered into Maple. First, enter $f := x \to x^2$ as a function.

```
> f:=x->x^2;
```
$$f := x \to x^2$$

Then enter Δx (here we use dx in the Maple session)

```
> dx:=(b-a)/n;
```
$$dx := \frac{b - a}{n}$$

The above sum is entered with Maple's **Sum** command.

```
> Sum(f(a+i*dx)*dx,i=1..n);
```
$$\sum_{i=1}^{n} \frac{\left(a + \frac{i\,(b-a)}{n}\right)^2 (b-a)}{n}$$

This sum can also be displayed without computing Δx and x_i by using the command **rightsum(f(x),x=a..b,n);** (this command requires the student package, **with(student);** that we loaded in the previous section).

The **Sum** command with an upper case S will display the sum without calculating its value, as illustrated above. This allows the sum to be checked for typing errors. The command **value(%);** can be added on the same line to evaluate this sum.

> **area:=Sum(f(a+i*dx)*dx,i=1..n); value(%);**

$$area := \sum_{i=1}^{n} \frac{\left(a + \dfrac{i(b-a)}{n}\right)^2 (b-a)}{n}$$

$$\frac{a^2(n+1)b}{n} - \frac{a^3(n+1)}{n} + \frac{ab^2(n+1)^2}{n^2} - \frac{ab^2(n+1)}{n^2}$$

$$-2\frac{a^2b(n+1)^2}{n^2} + 2\frac{a^2b(n+1)}{n^2} + \frac{a^3(n+1)^2}{n^2} - \frac{a^3(n+1)}{n^2}$$

$$+\frac{1}{3}\frac{b^3(n+1)^3}{n^3} - \frac{1}{2}\frac{b^3(n+1)^2}{n^3} + \frac{1}{6}\frac{b^3(n+1)}{n^3} - \frac{b^2a(n+1)^3}{n^3}$$

$$+\frac{3}{2}\frac{b^2a(n+1)^2}{n^3} - \frac{1}{2}\frac{b^2a(n+1)}{n^3} + \frac{ba^2(n+1)^3}{n^3}$$

$$-\frac{3}{2}\frac{ba^2(n+1)^2}{n^3} + \frac{1}{2}\frac{ba^2(n+1)}{n^3} - \frac{1}{3}\frac{a^3(n+1)^3}{n^3} + \frac{1}{2}\frac{a^3(n+1)^2}{n^3}$$

$$-\frac{1}{6}\frac{a^3(n+1)}{n^3} - \frac{a^2b}{n} + \frac{a^3}{n}$$

The lengthy Maple output represents the sum of the areas of the n rectangles that approximates the area under the graph of $f(x) = x^2$ from $x = a$ to $x = b$. To get the precise area under the graph, take the limit of this expression as n approaches infinity.

> **Limit(area,n=infinity); value(%);**

$$\lim_{n\to\infty} \sum_{i=1}^{n} \frac{\left(a + \dfrac{i(b-a)}{n}\right)^2 (b-a)}{n}$$

$$\frac{1}{3}b^3 - \frac{1}{3}a^3$$

Note that this expression is the same as

$$F(b) - F(a)$$

where $F(x) = \dfrac{x^3}{3}$. Also note that $F(x) = \dfrac{x^3}{3}$ is an antiderivative of $f(x) = x^2$. This is an illustration of the *Fundamental Theorem of Calculus*, which states that the definite integral $\int_a^b f(x)\,dx$ (defined in terms of the limit of the sum of areas of rectangles) is the same as $F(b) - F(a)$, where F is an antiderivative of f.

As a further illustration of the Fundamental Theorem, try repeating the above procedure with the functions $f(x) = x^3$ and $f(x) = x^4$. The only Maple

command that you must change is the one that involves the definition of the
function f. The other statements can be re-executed without change.

Maple has a command that integrates expressions. To integrate a function
such as $f := x \rightarrow x^2$, you must enter

> Int(f(x),x=a..b);

$$\int_a^b x^2 \, dx$$

Analogous to the **Sum** command, the **Int** command with an uppercase I displays
the integral without evaluating it, so that it can be checked for typing errors.
The integral can be evaluated by adding the command **value(%);**.

> Int(f(x),x=a..b); value(%);

$$\int_a^b x^2 \, dx$$

$$\frac{1}{3} b^3 - \frac{1}{3} a^3$$

The antiderivative of f (or indefinite integral) can also be evaluated.

> Int(f(x),x); value(%);

$$\int x^2 \, dx$$

$$\frac{1}{3} x^3$$

Note that Maple does not insert the constant of integration.

If f is defined as an expression rather than a function (via **f:=x^2; **), the
same syntax as above is used to integrate f, except that f is typed instead
of $f(x)$. That is, **Int(f,x=a..b);** will display the integral $\int_a^b x^2 \, dx$, and then
value(%) will evaluate it.

Maple cannot find an antiderivative for every expression. For example, try
integrating $\sqrt{x^5 + 1}$ with Maple. Some definite integrals have to be approximated by adding up rectangles or trapezoids or via some other more sophisticated technique. Maple makes it easy to evaluate an approximation to a definite
integral by using **evalf**. For example, the integral $\int_1^2 \sqrt{x^5 + 1} \, dx$ can be evaluated by entering

```
> Int(sqrt(x^5+1),x=1..2); evalf(%);
```

$$\int_1^2 \sqrt{x^5 + 1}\, dx$$

3.147104357

5.3 Integration by Substitution (Change of Variables)

Background Information: Read Section 5.5 in Stewart's **Calculus**.

Suppose we want to compute the integral $A = \int \dfrac{1}{\sqrt{4 + 9x^2}}\, dx$ by making a change of variables. Actually, Maple can integrate this without changing variables, but it's easier to understand a technique when it is demonstrated on an easy example.

The first thing to do is to identify a relation between the old variable x and a new variable. In this case, take

$$3x = 2\tan\theta$$

where the new variable is θ. Compute the differential of this relation and solve for dx. We now have

$$3\, dx = 2\sec^2\theta\, d\theta$$
$$dx = \frac{2}{3}\sec^2\theta\, d\theta$$

Next, substitute into the integral to get

$$A = \int \frac{2\sec^2\theta}{3\sqrt{4 + 4\tan^2\theta}}\, d\theta = \frac{1}{3}\int \sec\theta\, d\theta = \frac{1}{3}\ln|\sec\theta + \tan\theta| + C$$

Finally, substitute back to get

$$A = \frac{1}{3}\ln\left|\frac{\sqrt{4 + 9x^2}}{2} + \frac{3}{2}x\right| + C$$

To do this computation using Maple, we use the **changevar** command within the **student package**, which has three arguments: the first argument is the relation between the old and new variables; the second argument is the integral; and the third argument is the name of the new variable. So to compute $\int \dfrac{1}{\sqrt{4 + 9x^2}}\, dx$, load the student package, define the integral (using the **Int** command), execute the change of variables (using **changevar**), and evaluate the integral (using **value**).

```
> with(student):
> A:=Int(1/sqrt(4+9*x^2),x);
```

$$A := \int \frac{1}{\sqrt{4+9\,x^2}}\, dx$$

```
> changevar(3*x=2*tan(theta),A,theta); Atemp:=value(%);
```

$$\int \frac{1}{3} \frac{2+2\tan(\theta)^2}{\sqrt{4+4\tan(\theta)^2}}\, d\theta$$

$$Atemp := \frac{1}{3} \operatorname{arcsinh}(\tan(\theta))$$

Finally, to substitute back, solve for the new variable and use the **subs** command.

```
> subs(theta=arctan(3*x/2),Atemp); Avalue:=simplify(%);
> convert(%,ln);
```

$$\frac{1}{3} \operatorname{arcsinh}\left(\tan\left(\arctan\left(\frac{3}{2} x \right) \right) \right)$$

$$Avalue := \frac{1}{3} \operatorname{arcsinh}\left(\frac{3}{2} x \right)$$

$$\frac{1}{3} \ln(\frac{3}{2} x + \frac{1}{2} \sqrt{4+9\,x^2})$$

This is the answer up to an additive constant. Maple's first answer looks different from the one we obtained by hand, but the convert command shows that they are equivalent (Maple uses the formula $\sinh^{-1}(w) = \ln\left(\sqrt{w^2+1}+w\right)$ for the conversion). However, as with any indefinite integral, check the answer by differentiating the results. Thus, executing

```
> diff(Avalue,x); simplify(%);
```

$$\frac{1}{2} \frac{1}{\sqrt{1+\frac{9}{4}\,x^2}}$$

$$\frac{1}{\sqrt{4+9\,x^2}}$$

gives back the integrand of the original integrand in A.

The first two commands could have been combined as follows.

```
> changevar(3*x=2*tan(theta),Int(1/sqrt(4+9*x^2),x),theta);
```

but it is better to do this in two steps. Define A first; then use the **changevar** command. In this way the integral can be checked for typos.

If the integral becomes more complicated on applying the **changevar** command, go back and try a different substitution or a different integration trick.

You may ask why the **changevar** command is used when the commands

```
> A:=Int(1/sqrt(4+9*x^2),x); value(%);
```

will give the same result. The answer is that there are integrals that Maple cannot compute directly. Then an intelligent human-computer interaction may produce a result that Maple could not do on its own. For examples of such integrals, see the exercises.

5.4 Integration by Parts

Background Information: Read Section 8.1 in Stewart's **Calculus.**

Suppose we want to compute the integral $\int x \sin x \, dx$ by the method of integration by parts. The first thing to do is to identify u and dv. In this case take

$$u = x \text{ and } dv = \sin x \, dx$$

Then compute du and v. We then have

$$du = dx \text{ and } v = -\cos x$$

The integration by parts formula $\int u \, dv = uv - \int v \, du$ gives

$$\int x \sin x \, dx = -x \cos x + \int \cos x \, dx = -x \cos x + \sin x + C$$

To do this computation using Maple, use the **intparts** command, which has two arguments: the first argument is the integral and the second argument is the part of the integrand that will be taken as u. So, to compute $\int x \sin x \, dx$, define the integral (using the **Int** command), execute the integration by parts (using **intparts**), and evaluate the integral (using **value**).

```
> A:=Int(x*sin(x),x); Aparts:=intparts(A,x); value(%);
```

$$A := \int x \sin(x) \, dx$$

$$Aparts := -x \cos(x) - \int -\cos(x) \, dx$$

$$-x \cos(x) + \sin(x)$$

Again, we could have combined these commands as

```
> intparts(Int(x*sin(x),x),x); value(%);
```

but it is better to work the problem in two steps in order to verify that the integral is entered correctly.

If the integral becomes more complicated on applying the **intparts** command, go back and try a different u or a different integration technique.

5.5   Integration of Rational Functions by Partial Fractions

Background Information: See Section 8.4 in Stewart's **Calculus**.

The **convert** command used for partial fractions requires a rational function as its first argument, the keyword **parfrac** as the second argument, and the independent variable as the third argument. For example, to evaluate $\int \dfrac{x^2 - 3x + 1}{x^3 + x^2 - 2x}\, dx$, define the integrand

> f:=(x^2-3*x+1)/(x^3+x^2-2*x);
$$f := \frac{x^2 - 3\,x + 1}{x^3 + x^2 - 2\,x}$$

Find its partial fraction expansion

> fpar:=convert(f,parfrac,x);
$$fpar := -\frac{1}{2}\frac{1}{x} + \frac{11}{6}\frac{1}{x+2} - \frac{1}{3}\frac{1}{x-1}$$

Construct the integral and find its value

> Int(fpar,x); value(%);
$$\int -\frac{1}{2}\frac{1}{x} + \frac{11}{6}\frac{1}{x+2} - \frac{1}{3}\frac{1}{x-1}\, dx$$
$$-\frac{1}{2}\ln(x) + \frac{11}{6}\ln(x+2) - \frac{1}{3}\ln(x-1)$$

Note that Maple omits the *absolute value* in its answer. It simply gives *an* antiderivative!

5.6   Approximate Integration

Background Information: Read Section 8.7 in Stewart's **Calculus**.

In this section we demonstrate Maple's built-in capability for using the midpoint rule, the trapedoidal rule and Simpson's rule. We start by loading the **student** package.

```
> with(student):
```

The Maple commands **middlesum**, **trapezoid** and **simpson** can be found in this package. Let's use these commands to approximate

$$\int_0^1 \sin(x^2 + 1)dx$$

The first step is to define the function $f(x) = \sin(x^2 + 1)$.

```
> f:=x->sin(x^2+1);
```
$$f := x \rightarrow \sin(x^2 + 1)$$

The midpoint rule and trapedoidal rule are implemented below with $n = 10$.

```
> middlesum(f(x),x=0..Pi,10);
```
$$\frac{1}{10}\pi \left(\sum_{i=0}^{9} \sin(\frac{1}{100}(i + \frac{1}{2})^2 \pi^2 + 1) \right)$$

```
> m:=evalf(%);
```
$$m := .8977656549$$

```
> trapezoid(f(x),x=0..Pi,10);
```
$$\frac{1}{20}\pi \left(\sin(1) + 2 \left(\sum_{i=1}^{9} \sin(\frac{1}{100} i^2 \pi^2 + 1) \right) + \sin(\pi^2 + 1) \right)$$

```
> t:=evalf(%);
```
$$t := .8858261305$$

These results are fairly close. Let's compute Simpson's rule with $n = 20$.

```
> s:=evalf(simpson(f(x),x=0..Pi,20));
```
$$s := .8937858135$$

Notice that $s = \frac{1}{3}t + \frac{2}{3}m$, as stated in Stewart's section 5.8.

```
> 2/3*m+1/3*t-s;
```
$$-.1\,10^{-9}$$

OK, now let's use the error bound for Simpson's rule to determine a value of

n which can be used to approximate the integral above within 10^{-5}. Recall that the error estimate for approximating an integral of the form $\int_a^b f(x)dx$ using Simpson's rule is given by

$$|E_S| \leq \frac{K(b-a)^5}{180n^4}$$

where $\left|f^{(4)}(x)\right| \leq K$ for all $a \leq x \leq b$. Consequently, we need to investigate $f^{(4)}(x)$. We compute the derivative and create the plot below.

```
> d4f:=(D@@4)(f);
```
$$d4f := x \rightarrow 16\sin(x^2+1)\,x^4 - 48\cos(x^2+1)\,x^2 - 12\sin(x^2+1)$$

```
> plot(d4f(x),x=0..Pi);
```

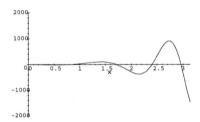

We can see from the plot that $\left|f^{(4)}(x)\right| \leq \left|f^{(4)}(\pi)\right|$. This value is defined as K below, and the equation

$$\frac{K\pi^5}{180n^4} = 10^{-5}$$

is solved for n.

```
> K:=abs(evalf(d4f(Pi)));
```
$$K := 1474.771668$$

```
> error_bound:=evalf(K*Pi^5/(180*n^4));
```
$$error\_bound := \frac{2507.273118}{n^4}$$

```
> fsolve(error_bound=0.00001,n=0..200);
```
$$125.8346980$$

Consequently, if we apply Simpson's rule with $n = 126$ then the result should have an error of no more than 10^{-5}.

```
> evalf(simpson(f(x),x=0..Pi,126));
                        .8934803287
```

Remark: Maple uses a much more sophisticated method for numerically approximating integrals. It might be interesting to compare Maple's approximation with the approximation from Simpson's rule for various values of n.

```
> evalf(int(f(x),x=0..Pi));
                        .8934801804
```

5.7   Summary

- Load **with(student)** package and use **leftbox** and **rightbox** to help visualize rectangles in Riemann sums.

- Use Maple to set up a Riemann sum over the interval from a to b using n rectangles.

```
> expr;                          # Enter expr, check for errors.
> f:=unapply(expr,x);            # Change to Maple function form.
> dx:=(b-a)/n;                   # Compute rectangle width.
> area:=Sum(f(a+i*dx)*dx,i=1..n); # Show Riemann sum.
> value(%);                      # Compute the Riemann sum.
```

- Use this Riemann sum to compute area.

```
> Limit(area,n=infinity);     # Show the limit.
> value(%);                   # Compute limit, which is area.
```

- Set up Riemann sums with **leftsum** and **rightsum** in the **with(student)** package.

- Know the **Int–value** syntax to compute both definite and indefinite integrals.

```
> Int(f(x),x);       # Shows indefinite integral to be calculated.
> value(%);          # Computes indefinite integral expression.
> Int(f(x),x=a..b);  # Shows definite integral to be computed.
> value(%);          # Computes definite integral.
```

- Recall that Maple adds no constant of integration.

- Maple will only integrate an expression. If f is a Maple function, it must be evaluated to $f(x)$ before integrating.

- The Maple commands **changevar**, **intparts**, and **convert/parfrac** can be used to help with integration by substitution, integration by parts and integration by partial fraction decomposition. Both **changevar** and **intparts** are contained within the **student** package. Sometimes these commands can be used to help Maple calculate complicated integrals in closed form.

- The Maple commands **middlesum**, **trapezoid** and **simpson** are also contained within the **student** package. These commands can be used to approximate a definite integral using the midpoint rule, the trapezoidal rule and Simpson's rule respectively.

5.8   Exercises

1. Load the student package via **with(student);**.

   (a) Use the **leftbox** command with 20 rectangles to display the Riemann sum approximation to

   $$\int_0^2 \sqrt{8 - x^3}\, dx$$

   Does this set of rectangles over-approximate or under-approximate the integral?

   (b) Set up and evaluate the Riemann sum represented by the **leftbox** command for $n = 20, 40$, and 80 rectangles.

   (c) Compare the values of the Riemann sums in (b) to the value obtained by using Maple's **Int** command followed by **evalf(");** as done in the text. As the number of rectangles doubles, what happens to the difference between the value of the Riemann sum and the value of the integral?

2. Repeat Exercise 1 using the **rightbox** command (along with setting up and evaluating the corresponding Riemann sums for $n = 20, 40$, and 80 rectangles).

3. Repeat Exercises 1 and 2 with the function $f(x) = \dfrac{x^2 - 1}{x^2 + 1}$.

4. Use Maple to evaluate the following integrals.

   (a) $\displaystyle\int_1^3 x\sqrt{x^2 + 1}\, dx$

(b) $\int \cos^4(x)\,dx$

(c) $\int \sin^6(x)\,dx$

5. There are two ways to evaluate the integral $\int \sec^2(x)\tan(x)\,dx$ by u-substitution. The first way is to let $u = \sec(x)$ and the second is to let $u = \tan(x)$. Based on Maple's answer to this integral, which one is it using? These two methods of substitution lead to apparently different answers. Are they, in fact, the same? If they are different, then how do you explain the fact that they are answers to the same problem?

6. Express each of the following limits as an integral and then use Maple to evaluate. (The first two integrals are provided for you.)

(a) $\displaystyle\lim_{n\to\infty}\sum_{i=1}^{n}\frac{1}{n}\sqrt{1+\frac{i^2}{n^2}} = \int_0^1 \sqrt{1+x^2}\,dx$

(b) $\displaystyle\lim_{n\to\infty}\sum_{i=1}^{n}\frac{2}{n}\sqrt{1+\frac{4i^2}{n^2}} = \int_0^2 \sqrt{1+x^2}\,dx$

(c) $\displaystyle\lim_{n\to\infty}\sum_{i=1}^{n}\frac{\pi}{n}\sin^2\left(\frac{i\pi}{n}\right)$

(d) $\displaystyle\lim_{n\to\infty}\sum_{i=1}^{n}\frac{2}{n}\left(\left(1+\frac{2i}{n}\right)^3 + 3\left(1+\frac{2i}{n}\right)\right)$

7. *Area of Texas.* From Exercise 25 in Chapter 2, the northern and southern boundaries of the state of Texas are given by the following data.

```
> north:=[[0,0],[3,0],[3,4.5],[6,4.5],[6,2.2],[7,2.1],[8,1.8],
> [9,1.9],[10,1.8],[11,1.7],[11,-2.2]];
> south:=[[0,0],[1,-1.1],[2,-2.5],[3,-2.9],[4,-2.3],[5,-2.8],
> [6,-4.4],[7,-5.8],[8,-6.1],[9,-3.3],[10,-2.8],[11,-2.2]];
```

(Enter these in Maple.) Here, the origin is the western corner of Texas (near El Paso) and the x-axis is the extension of the east-west border between New Mexico and Texas. Each unit represents approximately 69 miles. Use these data to approximate the area of Texas by using a Riemann sum formed from rectangles whose widths are 1 unit and whose heights are determined by the second coordinates of the data. *Hint:* To refer to the entries in a list, use brackets []. For example, **south[3]** refers to the point $[2, -2.5]$ and **south[3][2]** refers to the second entry of this point, -2.5. So to (symbolically) sum up all second entries of the points on this list, enter **Sum(south[i][2],i=1..11);**. Then, to actually compute the sum, enter **value(");**.

8. Evaluate the integral $\int \dfrac{x^2 \tan^{-1}(x)}{(1+x^2)^2}\, dx$. First try the **value** command by itself. Then use the **changevar, value, subs,** and **simplify** commands. Be sure to check your answer.

9. Evaluate the integral $\int x \ln\left(x + \sqrt{1+x^2}\right) dx$. First try the **value** command by itself. Then use the **intparts** and **value** commands with $u = x$. Finally, use the **intparts** and **value** commands with $u = \ln\left(x + \sqrt{1+x^2}\right)$. Be sure to check the answer.

10. Find the partial fraction expansion for $f = \dfrac{x^8 + 2x - 1}{(x-1)^3(x^2+3)^2}$ and use it to evaluate $\int \dfrac{x^8 + 2x - 1}{(x-1)^3(x^2+3)^2}\, dx$.

Use Maple to compute the following integrals.

11. $\int \dfrac{x}{x^6 + 1}\, dx$

12. $\int \dfrac{\tan\left(\sqrt{x}\right)}{2\sqrt{x}}\, dx$

13. $\int \sin^4(x) \cos^2(x)\, dx$

14. $\int_{-1}^{3} \dfrac{x^3 + x}{\sqrt{1+x^2}}\, dx$

15. $\int_{-\pi/4}^{\pi/4} \sqrt{\tan^2(x) + 1}\, dx$

16. Use each of the midpoint rule, the trapezoidal rule and Simpson's rule with $n = 20, 40, 80$ to approximate the integral $\int_0^2 \sqrt{8 - x^3}\, dx$.

17. Use the error estimate for Simpson's rule to determine a value for n so that if Simpson's rule is used to approximate the integral $\int_0^{1.5} \sqrt{8 - x^3}\, dx$ then the value is accurate to 5 decimal places.

Chapter 6

Applications of Integration

Background Information: Read Sections 6.1, 6.2 and 6.3 in Stewart's **Calculus.**

The first section in this chapter uses Maple to compute areas bounded by curves. The second section discusses the use of Maple in the computation of volumes of revolution. We conclude this chapter with an application of integration to the approximation of functions by constants and cosines.

6.1   Area

Background Information: Read Section 6.1 in Stewart's **Calculus.**

We present two examples in this section. In the first example, the area between a curve and the x-axis is computed. The area between two curves is considered in the second example.

Example 1. Find the area that lies between the curve given by

$$f(x) = -0.128x^3 + 1.728x^2 - 5.376x + 2.864$$

and the x-axis.

Solution: Start by inputting an expression for f into Maple.

```
> f:=-0.128*x^3+1.728*x^2-5.376*x+2.864;
                f := -.128 x^3 + 1.728 x^2 - 5.376 x + 2.864
```

Now plot f over an interval that shows all points where f crosses the x-axis (in this case, the interval $-2 \le x \le 10$ will do).

```
> plot(f,x=-2..10);
```

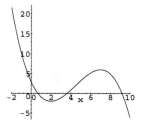

The graph of f crosses the x-axis at three points—near 0, between 3 and 4, and between 9 and 10. Use **fsolve** to find these roots and assign them to a variable, say rt.

> **rt:=fsolve(f=0,x);**

$$rt := .6697777844, 3.631759112, 9.198463104$$

Note that all the roots can be found with **fsolve** without specifying a range for x, because f is a polynomial. The variable rt is a Maple sequence and we can refer to each of its elements by typing $rt[1]$, $rt[2]$, etc. From the plot, it is evident that the graph of f is below the x-axis between $rt[1]$ and $rt[2]$, and the graph of f is above the x-axis between $rt[2]$ and $rt[3]$. Therefore, the following integral must be computed

$$-\int_{rt[1]}^{rt[2]} f\,dx + \int_{rt[2]}^{rt[3]} f\,dx$$

In Maple we enter

> **-Int(f,x=rt[1]..rt[2])+Int(f,x=rt[2]..rt[3]);**

$$-\int_{.6697777844}^{3.631759112} -.128\,x^3 + 1.728\,x^2 - 5.376\,x + 2.864\,dx$$
$$+\int_{3.631759112}^{9.198463104} -.128\,x^3 + 1.728\,x^2 - 5.376\,x + 2.864\,dx$$

As mentioned in the last chapter, the **Int** command (with an uppercase I) displays the integral without computing its value so that it can be checked for typing errors. Now use the **value** command to evaluate this integral.

> **value(%);**

$$25.05016802$$

The area of interest is about 25.05 square units.

Example 2. Compute the area between the graph of f (defined above) and the graph of

$$g(x) = 0.08x^3 - 0.84x^2 + 1.44x + 4.32$$

Solution: Once again, we start by defining g as an expression in Maple.

```
> g:=0.08*x^3-0.84*x^2+1.44*x+4.32;
```
$$g := .08\,x^3 - .84\,x^2 + 1.44\,x + 4.32$$

Now plot f and g on the same coordinate axes with the command

```
> plot({f,g},x=-2..10);
```

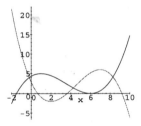

The plot shows three points of intersection, which can be found with **fsolve** (once again **fsolve** will find all of the solutions since f and g are polynomials).

```
> rt:=fsolve(f=g,x);
```
$$rt := -.1985269946, 4.251804404, 8.292876437$$

The graph of f is below the graph of g between *rt[1]* and *rt[2]*, and the graph of g is below the graph of f between *rt[2]* and *rt[3]*. Therefore, the area between the graph of f and g is computed by the following integral

$$\int_{rt[1]}^{rt[2]} (g - f)\, dx + \int_{rt[2]}^{rt[3]} (f - g)\, dx$$

In Maple this is entered as

```
> Int(g-f,x=rt[1]..rt[2])+Int(f-g,x=rt[2]..rt[3]);
```

$$\int_{-.1985269946}^{4.251804404} .208\,x^3 - 2.568\,x^2 + 6.816\,x + 1.456\,dx$$

$$+ \int_{4.251804404}^{8.292876437} -.208\,x^3 + 2.568\,x^2 - 6.816\,x - 1.456\,dx$$

Following this command by **value(%);** yields the answer, 33.95029006.

6.2   Volume

Background Information: Read Sections 6.2 and 6.3 in Stewart's **Calculus.**
   To calculate a volume by slicing, the first step is to derive an expression that represents the cross sectional area. Then this expression is integrated over the appropriate interval.

Example 1.   Consider the region that is bounded above by the curve $y = -x^2 + 5x - 2$ and below by the line $y = x$. Find the volume of the solid that is obtained by revolving this region about the x-axis.
Solution. First, define the expressions $f := -x^2 + 5x - 2$ and $g := x$ in Maple and graph these expressions with a plot command.

```
> f:=-x^2+5*x-2;
> g:=x;
> plot({f,g},x=0..4);
```

$$f := -x^2 + 5x - 2$$

$$g := x$$

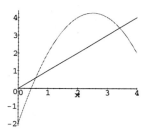

   From the plot, it is evident that f and g cross at two points: one with an x-coordinate between $x = 0$ and $x = 1$ and the other with an x-coordinate between $x = 3$ and $x = 4$. These roots can be determined with the following **fsolve** commands and assigned to the variables a and b.

> a:=fsolve(f=g,x,x=0..1); b:=fsolve(f=g,x,x=3..4);

$$a := .5857864376$$

$$b := 3.414213562$$

Here, a range for the **fsolve** command is given because we want a to designate the lower limit and b to designate the upper limit. The cross section of this solid is in the shape of a *washer* whose outside radius is given by f and whose inside radius is given by g.

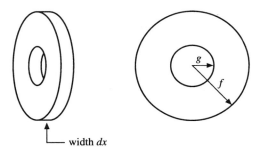

The area of the cross section is given by $\pi(f^2-g^2)$. The total volume of this solid is obtained by summing (integrating) this expression over the interval $a \leq x \leq b$. This is easily accomplished in Maple.

> Int(Pi*(f^2-g^2),x=a..b); value(%);

$$\int_{.5857864376}^{3.414213562} \pi\left(\left(-x^2 + 5x - 2\right)^2 - x^2\right)\, dx$$

$$66.34705188$$

Recall that the first command displays the integral, while the second command computes its value.

Example 2. Find the volume of the solid obtained by revolving this region about the y-axis.

Solution. In this case, finding the area of the cross section (which is a washer perpendicular to the y-axis) is more difficult (try this as a challenge). An easier method is to calculate the volume by the method of cylindrical shells. The idea is to slice the volume by shells whose radius, height, and thickness are given by

x, $f - g$, and dx, respectively.

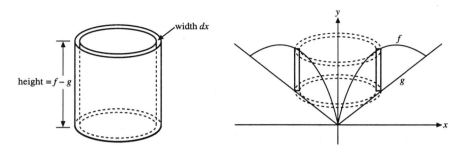

The volume of this shell is given by $2\pi x(f - g)\, dx$. The total volume of this solid is obtained by summing (integrating) this expression over the interval $a \le x \le b$.

> **Int(2*Pi*x*(f-g),x=a..b); value(%);**
$$\int_{.5857864376}^{3.414213562} 2\,\pi\,x\left(-x^2 + 4\,x - 2\right) dx$$

$$47.39075130$$

6.3   An Introduction to Fourier Series - Cosine Expansions

Background Information: This is an optional section. Read Chapter 5 and Section 8.2 in Stewart's **Calculus**.

One of the interesting applications of integration is the approximation of functions. We will demonstrate in this section that smooth functions can be approximated with trigonometric functions. More appropriately, we will demonstrate that a function on an interval of the form $[0, L]$ can be approximated by the functions

$$1,\ \cos\left(\frac{\pi}{L}x\right),\ \cos\left(\frac{2\pi}{L}x\right),\ \ldots,\ \cos\left(\frac{n\pi}{L}x\right) \tag{6.1}$$

by taking n sufficiently large.

To do this, we need to learn about the properties of the functions above. First, let's notice that

$$\int_0^L \cos\left(\frac{k\pi}{L}x\right) dx = \frac{\sin k\pi}{k\pi}L = 0 \text{ for all } k = 1, 2, 3, \ldots$$

Furthermore, if m and k are positive integers with $m \ne k$, then

$$\int_0^L \cos\left(\frac{k\pi}{L}x\right) \cos\left(\frac{m\pi}{L}x\right) dx = 0$$

(do the integration and see for yourself!!) Finally, if k is a positive integer then

$$\int_0^L \cos^2\left(\frac{k\pi}{L}x\right) dx = \frac{1}{2}L\frac{\cos k\pi \sin k\pi + k\pi}{k\pi} = \frac{L}{2}$$

and

$$\int_0^L 1 dx = L.$$

Let's use this information to try and find a way to approximate a function $f(x)$ in the form

$$f(x) \approx a_0 + a_1 \cos\left(\frac{\pi}{L}x\right) + ... + a_n \cos\left(\frac{n\pi}{L}x\right) \tag{6.2}$$

on the interval $[0, L]$. In order to determine the coefficient a_0, we simply integrate both sides of (6.2) above and obtain

$$\int_0^L f(x)dx = a_0 L$$

(since all the other terms integrate to zero). This gives,

$$a_0 = \frac{1}{L}\int_0^L f(x)dx \tag{6.3}$$

(Note that a_0 is the average value of $f(x)$ on the interval $[0, L]$.) Furthermore, we can obtain a formula for a_1 by multiplying both sides in (6.2) by $\cos(\frac{\pi}{L}x)$ and integrating. This yields

$$\int_0^L f(x)\cos(\frac{\pi}{L}x)dx = \int_0^L a_1 \cos^2(\frac{\pi}{L}x)dx$$

since all the other terms integrate to zero. Consequently, we have

$$a_1 = \frac{2}{L}\int_0^L f(x)\cos(\frac{\pi}{L}x)dx.$$

a_2 can be found in a similar manner by multiplying both sides of (6.2) by $\cos(\frac{2\pi}{L}x)$ and integrating. In this case we obtain

$$a_2 = \frac{2}{L}\int_0^L f(x)\cos(\frac{2\pi}{L}x)dx.$$

In general, we obtain

$$a_k = \frac{2}{L}\int_0^L f(x)\cos(\frac{k\pi}{L}x)dx, \quad k = 1, 2, ..., n \tag{6.4}$$

The terms a_0, a_1, ... a_n are called the **Fourier Cosine coefficients** of the function $f(x)$ on the interval $[0, L]$.

Definition: Let $f : [0, L] \rightarrow R$ be a smooth function. If n is a positive integer then the n-th order Fourier Cosine expansion for $f(x)$ on the interval $[0, L]$ is given by

$$a_0 + \sum_{k=1}^{n} a_k \cos(\frac{k\pi}{L}x)$$

where a_0 and a_k are given by the formulas in (6.3) and (6.4) respectively.

Example 1. Compute the 10-th order Fourier Cosine expansion for the function $f(x) = x^2 - 1$ on the interval $[0, 3]$.

Solution: Using the formulas above, we can input f and create the Fourier coefficients as follows:

```
> f:=x^2-1:
> 1/3*Int(f,x=0..3);
> a0:=value(%);
```

$$\frac{1}{3}\int_0^3 x^2 - 1\,dx$$

$$a0 := 2$$

```
> assume(k,integer):
> ak:=2/3*Int(f*cos(k*Pi*x/3),x=0..3);
> ak:=value(%);
```

$$ak := 36\frac{(-1)^{k\tilde{}}}{k\tilde{}^2\,\pi^2}$$

Consequently, we can define the 10-th order Fourier Cosine expansion for the function $f(x) = x^2 - 1$ on the interval $[0, 3]$ and plot this expression against $f(x)$ on the interval $[0, 3]$ to see the quality of our approximation.

```
> fcos:=a0+sum(ak*cos(k*Pi*x/3),k=1..10):
> plot({fcos,f},x=0..3);
```

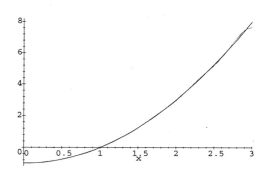

It looks as though the approximation is pretty good, with the worst part of it being at the endpoint $x = 3$. The reason for the poor behavior there is simple. All of the functions in (6.1) (with $L = 3$) have a zero derivative at $x = 3$. Since the function $f(x) = x^2 - 1$ does not have this property, we can expect the approximating Fourier expansion to have some trouble near $x = 3$. Notice that there does not seem to be a problem at $x = 0$. This is because $f'(0) = 0$, and all of the functions in (6.1) have this property. In addition, notice what happens when we look at the graph of $f(x)$ versus this Fourier Cosine expansion on a larger interval. In particular, let's try the interval $[-3, 3]$. Here we have

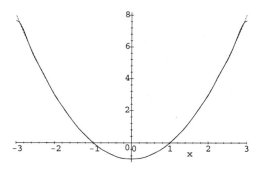

and the agreement is quite good. If you are wondering whether this will always happen, then the answer is a definite NO! The reason it happens here is because $f(x) = x^2 - 1$ is an even function, and all of the functions in (6.1) are also even functions! Finally, let's look at a slightly larger interval.

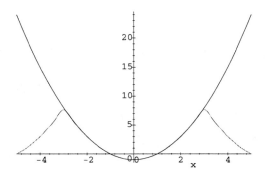

Notice that the Fourier Cosine expansion falls away quickly outside the interval $[-3, 3]$. You could say that the expansion did what it was asked to do. However, in actuality, the fall off takes place because all of the functions in (6.1) are periodic with period 6 (with the choice $L = 3$) and $f(x)$ is not a periodic function.

Example 2: Let's see what happens if we attempt to approximate a function which is not smooth. In fact, let's try one which is not continuous. For example, consider the function $g(x) = \begin{cases} x & \text{if} & x < 1 \\ 3 - x & \text{if} & x > 1 \end{cases}$. This function can be input in Maple and graphed as follows:

```
> g:=piecewise(x<1,x,3-x);
> plot(g,x=0..3,style=point);
```

$$g := \begin{cases} x & x < 1 \\ 3 - x & \textit{otherwise} \end{cases}$$

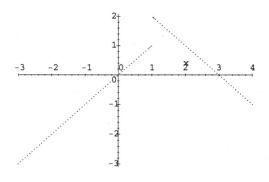

You might notice that we have given a "point-plot" to keep Maple from connecting the dots. This way we can actually see the jump discontinuity at $x = 1$. Let's try to approximate the function $g(x)$ on the interval $[0, 2]$ with a Fourier Cosine expansion of order 20.

Solution: First, we need to generate some coefficients.

```
> b0:=1/2*int(g,x=0..2);
```

$$b0 := 1$$

```
> bk:=int(g*cos(k*Pi*x/2),x=0..2);
```

$$bk := -2\,\frac{2\,(-1)^{k\tilde{}} - 4\cos(\frac{1}{2}\,k\tilde{}\,\pi) + k\tilde{}\,\pi\sin(\frac{1}{2}\,k\tilde{}\,\pi)}{k\tilde{}^2\,\pi^2} - \frac{4}{k\tilde{}^2\,\pi^2}$$

Consequently, we can create the Fourier Cosine expansion of order 20 and graph the expansion against $g(x)$ on the interval $[0, 2]$.

```
> gcos:=b0+sum(bk*cos(k*Pi*x/2),k=1..20):
> plot({g,gcos},x=0..2);
```

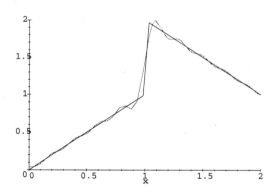

You should observe that the Fourier Cosine expansion does a very good job of approximating $g(x)$ so long as we stay away from the point of discontinuity at $x = 1$. At this point, it appears as though the Fourier Cosine expansion gives essentially the average of what $g(x)$ is trying to do from the left of $x = 1$ and what $g(x)$ is trying to do from the right of $x = 1$. That is, the average of 1 and 2; i.e. $\frac{3}{2}$.

```
> evalf(subs(x=1,gcos));
```

$$1.479802392$$

That looks close! In fact, it is generally the case that as you increase the order of the Fourier Cosine expansion, it's value at a jump discontinuity will converge to the average of the right and left hand limits at the discontinuity.

6.4   Summary

- Use **fsolve** to find successive points of intersection of the graphs of Maple expressions f and g. Use **Int–value** syntax applied to the difference $f - g$, followed by the absolute value function, to evaluate the area between successive crossings. Add areas between crossings.

- Use **Int–value** syntax to compute volumes of solids of revolution by slicing solids into annular washers or cylindrical shells.

- Use **int** and **sum** commands to create Fourier cosines series expansions.

6.5   Exercises

1. Find the area of the region that is bounded above by the curves $y = 7 \ln (x)$ and $y = 4 - x^3 - x$ and below by the x-axis.

2. Consider the region in the previous exercise.

   (a) Find the volume of the solid obtained by revolving this region about the x-axis.

   (b) Find the volume of the solid obtained by revolving this region about the y-axis.

   (c) Find the volume of the solid obtained by revolving this region about the line $x = 4$.

   (d) Find the volume of the solid obtained by revolving this region about the line $y = 4$.

3. Consider the solid obtained by revolving the region bounded by the graph of $y = 16 - x^2$ and the positive coordinate axes about the x-axis. Use the **Sum** and **value** commands to approximate the volume of this solid by summing the volumes of 10, 20, 100, and 1000 inscribed shells of equal thickness. Now repeat with circumscribed shells.

4. Compute the volume of the solid in Exercise 3 by integration and compare your answers.

5. This problem is a continuation of Exercise 11 in Chapter 4. In that exercise, the northern boundary of a plot of land is described by the following table (each unit represents 100 feet).

x-value	0.0	1.0	2.0	3.0	4.0
y-value	0.0	2.1	3.2	2.4	1.7

where the x-axis represents the (flat) southern boundary with the origin located at its western corner. Use these data to compute an approximation to the area of this plot of land using rectangles.

6. Now do Exercise 11 from Chapter 4 or load the answer from a file if you have already done that exercise. Use the quadratic and cubic functions described therein to evaluate the area of this plot of land.

7. *The volume of a JELL-O*® *mold.* Consider the region bounded below by the x-axis and above by the parabola that passes through the points $(8, 0)$, $(10, 4)$, and $(12, 0)$.

   (a) Find the volume of the solid obtained by revolving this region about the y-axis.

   (b) Find the volume of the solid obtained by revolving this region about the line $x = 2$.

8. In this exercise, you are asked to calculate the approximate volume of a bowl of depth 3 feet with circular cross sections. The measurements of the radius of the cross sections versus depth are given in the following table (in feet).

Depth	0.00	0.50	1.00	1.50	2.00	2.50	3.00
Radius	2.00	1.85	1.60	1.40	1.05	0.55	0.00

   Find an approximate value of the volume of this bowl by adding the volumes of the shells of width 0.5 feet and radius given by the values in the table above.

9. Find a third degree polynomial whose graph contains every other data point in the table of the previous exercise. Use this polynomial to compute an approximate volume for the bowl. Compare your answer to that of the previous exercise. *Hint:* Think of the x-axis as representing depth, and enter $p(x) = ax^3 + bx^2 + cx + d$ as a function in Maple. Then solve the four equations $p(0) = 2$, $p(1) = 1.6$, $p(2) = 1.05$, and $p(3) = 0$ for the unknowns a, b, c, and d. Graph this polynomial to see if its graph contains these data points.

10. A circular doughnut is formed by revolving a circle of radius r centered at $x = a$, $y = 0$ about the y-axis ($a > r$). Your job is to find the formula for the volume of this doughnut.

   (a) Find the equation of the circle described above.

   (b) Review the cylindrical shell technique for finding volumes and set up the integral for the volume of this doughnut.

   (c) Use Maple's **Int** and **value** commands to evaluate this integral.

(d) Evaluate the volume in the special case where $a = 3$ and $r = 1$.

(e) To see a three-dimensional picture of this doughnut, issue the following Maple commands.

```
> x:=cos(s)*(3+cos(t)); y:=sin(t); z:=sin(s)*(3+cos(t));
```

See if you can figure out why these three equations parameterize the doughnut as the parameters s and t vary from 0 to 2π. Then type

```
> with(plots): plot3d([x,y,z],s=0..2*Pi,t=0..2*Pi);
```

11. Generate the 10-th and 15-th order Fourier Cosine expansions for the function $f(x) = \sin(x)$ on the interval $[0, 3]$. Plot each of these against the function $f(x)$.

12. Repeat the previous exercise with $f(x) = 2 - x^2$ on the interval $[0, 3]$.

13. Repeat the previous exercise with $f(x) = \begin{cases} 2 - x & \text{if } x < 2 \\ x^2 - 1 & \text{if } x > 2 \end{cases}$ on the interval $[0, 3]$ with Fourier Cosine expansions of order 20 and 30. In each case, check the value of your Cosine expansion at the point of jump discontinuity. Do you observe the averaging property?

14. This exercise shows you how to obtain a Fourier Cosine expansion of order n on an arbitrary interval $[a, b]$. Show that if you want an approximation of the form

$$f(x) \approx a_0 + a_1 \cos\left(\frac{\pi}{b-a}(x - a)\right) + \ldots + a_n \cos\left(\frac{n\pi}{b-a}(x - a)\right)$$

then the Fourier Cosine coefficients should be

$$a_0 = \frac{1}{b-a} \int_a^b f(x)dx$$

and

$$a_k = \frac{2}{b-a} \int_a^b f(x) \cos\left(\frac{k\pi}{L}(x - a)\right) dx \quad \text{for } k = 1, 2, \ldots, n.$$

15. Use the information in exercise 14 to generate the 10-th order Fourier Cosine expansion for the function $f(x) = 10 - 15x - x^2$ on the interval $[3, 5]$. Plot the Cosine expansion against the function $f(x)$.

16. Repeat exercise 15 using the 20-th order Fourier Cosine expansion.

17. Repeat exercises 15 and 16 with the function $f(x) = \sin(x) + x$.

Chapter 7

Differential Equations

Background Information: Read Chapter 10 in Stewart's **Calculus.**

This chapter focusses on the use of Maple commands to analyze differential equations. We use Maple's **dsolve** command to find explicit solutions to various first and second order differential equations. Maple is very good at finding explicit solutions for differential equations—when they can be found. However, solutions to many of the important differential equations cannot be found in closed form. In this case, the **numeric** option of **dsolve** can be used to find approximate numerical solutions to differential equations. In addition, the **dfieldplot** command can be used to create direction fields, and the **phaseportrait** command can be used to create phase portraits.

7.1 Explicit Solutions

Background Information: Read Section 10.3 in Stewart's **Calculus.**

To solve the differential equation $y' + 5y = 2t$, input

```
> eq1:=diff(y(t),t)+5*y(t)=2*t; dsolve(eq1,y(t));
```
$$eq1 := \left(\frac{\partial}{\partial t}\, \mathrm{y}(t)\right) + 5\,\mathrm{y}(t) = 2\,t$$
$$\mathrm{y}(t) = \frac{2}{5}\,t - \frac{2}{25} + \mathrm{e}^{(-5\,t)}\,\_C1$$

Note that the symbol *y(t)* in Maple's output is neither an expression nor a function. It is a notational device used by Maple. Also notice the presence of the term *_C1*. This is Maple's notation for the arbitrary constant that occurs in the general solution of a first order differential equation.

An initial condition such as $y(1) = 2$ can be imposed in the following manner.

> dsolve({eq1,y(1)=2},y(t));

$$y(t) = \frac{2}{5}t - \frac{2}{25} + \frac{42}{25}\frac{e^{(-5t)}}{e^{(-5)}}$$

Note that the equation and the initial condition are enclosed in curly braces. To see a plot of the solution, the following Maple commands can be used:

> ysol:=rhs(dsolve({eq1,y(1)=2},y(t))):
> plot(ysol,t=0..3);

The command **rhs** refers to the right hand side of an equation.

The commands to solve a second (or higher) order differential equation are similar to the above. To find the general solution to the second order differential equation

$$\frac{d^2y}{dt^2} + 6\frac{dy}{dt} - 7y = 1$$

enter the commands

> y:='y'; eq2:=diff(y(t),t$2)+6*diff(y(t),t)-7*y(t)=1;
> dsolve(eq2,y(t));

$$y := y$$

$$eq2 := \left(\frac{\partial^2}{\partial t^2}y(t)\right) + 6\left(\frac{\partial}{\partial t}y(t)\right) - 7y(t) = 1$$

$$y(t) = -\frac{1}{7} + \_C1\,e^t + \_C2\,e^{(-7t)}$$

Notice that there are two arbitrary constants in the solution, $\_C1$ and $\_C2$. To specify initial conditions such as $y(1) = 3$ and $y'(1) = 2$, enter

> dsolve({eq2, y(1)=3, D(y)(1)=2},y(t));

$$y(t) = -\frac{1}{7} + 3\frac{e^t}{e} + \frac{1}{7}\frac{e^{(-7t)}}{e^{(-7)}}$$

Note that the **diff** command cannot be used to specify an initial condition in **dsolve**.

7.2   Direction Fields

Background Information: Read Section 10.2 in Stewart's **Calculus.**

A useful geometric device that aids in understanding the behavior of solutions to a first order differential equation is the direction field associated with the differential equation. Maple has the ability to easily plot this field. First load the Maple package **DEtools**.

```
> with(DEtools):
```

Suppose we want to investigate the solutions to the differential equation $y' = x \sin(y)$ by first examining its direction field.

```
> dfieldplot(diff(y(x),x)=x*sin(y(x)),y(x),x=-2..2,y=-1..1);
```

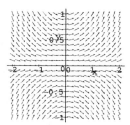

To graph a solution to this differential equation on top of the direction field, the following procedure is used. First load the **plots** package, and then give names to the objects you wish to plot.

```
> with(plots):
> plot1:=dfieldplot(diff(y(x),x)=x*sin(y(x)),y(x),x=-2..2,y=-1..1):
> dsolve({diff(y(x),x)=x*sin(y(x)),y(0)=.25},y(x));
> plot2:=plot(rhs("),x=-2..2):
> display([plot1,plot2]);
```

$$y(x) = 2\arctan\left(e^{\left(1/2\,x^2 - 2.074214136\right)}\right)$$

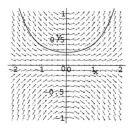

Notice how the solution follows the direction field. This is due to the fact that the tangent line to the solution is parallel to the line given by the direction field. In fact, it is this idea that gives rise to Euler's method, which is an easily implemented numerical algorithm for calculating approximate solutions to differential equations.

7.3   Numerical Solutions

Background Information: The material in this section is optional. However, since many nonlinear differential equations are virtually impossible to solve (in closed form), students should find the information in this section very useful. Read Section 10.2 in Stewart's **Calculus.**

Maple can find numerical approximations to differential equations. The algorithm that Maple uses is a Fehlberg fourth-fifth order Runge-Kutta method, which is more sophisticated than Euler's method. For example, suppose we want to solve $y' + \sin(y^2) = 1$, $y(0) = 1$. If we proceed as in the previous section, we get the following.

> eq3:=diff(y(t),t)+sin(y(t)^2)=1; dsolve({eq3,y(0)=1},y(t));

$$eq3 := \left(\frac{\partial}{\partial t}\, y(t)\right) + \sin\left(y(t)^2\right) = 1$$

```
Error, (in solve/sumint)
cannot solve for variables used in unevaluated sum/ints
```

Maple is telling us that it cannot evaluate some integral in closed form. To obtain an approximate solution using **dsolve**, the **numeric** option with an initial condition must be specified. Maple's output is then a procedure, which can be used in a manner similar to a Maple function. (For details on procedures, see Chapter 10.)

> dsolve({eq3,y(0)=1},y(t),numeric);

```
proc(rkf45_x) ... end
```

If we want to use this procedure to find approximate values for the solution to the differential equation at various values of t, the easiest way to do this is to name it. So go back and assign a name to the previous input line.

> f:=dsolve({eq3,y(0)=1},y(t),numeric);

```
f := proc(rkf45_x) ... end
```

Let's evaluate the solution at $t = 0$ and $t = 0.5$.

> f(0); f(0.5);

$$[t = 0, \mathrm{y}(t) = 1.]$$

$$[t = .5, \mathrm{y}(t) = 1.062031861131322]$$

To plot the solution we use Maple's **odeplot** command.

> with(plots): odeplot(f,[t,y(t)],0..4);

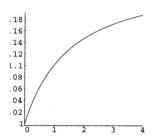

Note. The variables in the list $[t, y(t)]$ must be the same symbols for the independent and dependent variables that were used when the procedure f was defined.

7.4 Systems of Differential Equations

Background Information: Read Section 10.7 in Stewart's **Calculus.**

One of the nicest features of Maple is its ability to analyze systems of differential equations. This section demonstrates the use of several of Maple's tools which can be used to analyzing systems of the form

$$\begin{cases} x'(t) = f(x(t), y(t)) \\ y'(t) = g(x(t), y(t)) \end{cases}$$

We start by loading the **DEtools** package.

> with(DEtools):

We will focus on the **dfieldplot** and **phaseportrait** commands. The **dfieldplot** command can be used to create direction field plots for first order systems of equations. For example, consider the Lotka-Volterra model given by

$$\begin{cases} x'(t) = x(t)\,(1 - y(t)) \\ y'(t) = 0.3y(t)\,(x(t) - 1) \end{cases} \tag{7.1}$$

We can give a direction field plot of this system as follows.

```
> de1:=diff(x(t),t)=x(t)*(1-y(t)):
> de2:=diff(y(t),t)=0.3*y(t)*(x(t)-1):
> dfieldplot([de1,de2],[x(t),y(t)],t=0..1,x=0..2,y=0..2);
```

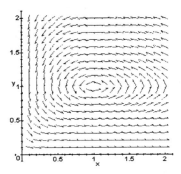

We can use the **phaseportrait** command to plot the direction field along with some phase trajectories (solution curves of (7.1)). Note the manner in which the two sets of initial data are specified below.

```
> phaseportrait([de1,de2],[x(t),y(t)],t=0..15,[[x(0)=1.5,y(0)=1.5],
> [x(0)=2,y(0)=2]],stepsize=0.1);
```

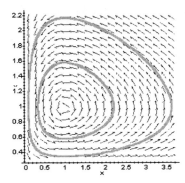

We conclude this section by demonstrating how Maple can be used to generate numerical approximations solutions to the system of differential equations given in (7.1). Suppose we want to obtain an approximation to this system corresponding to the initial data $x(0) = 1.5$ and $y(0) = 1.5$. We start by using the **numeric** option of the **dsolve** command to create a procedure.

```
> soln := dsolve({de1,de2,x(0)=1.5,y(0)=1.5},{x(t),y(t)}, type=numeric,
>            output=listprocedure,abserr=0.001):
```

We can use this procedure to obtain approximations for our solution at any time value. For example, the approximate values at $t = 1$ and $t = 2$ are shown below.

```
> soln(1);
```
$$[t(1) = 1,\ y(t)(1) = 1.571488025594804,\ x(t)(1) = .8563002959872530]$$

```
> soln(2);
```
$$[t(2) = 2,\ y(t)(2) = 1.420438434537689,\ x(t)(2) = .5155576279930636]$$

We can also force Maple to only extract the portions that we are interested in. If we only want to consider the approximate solution values for $x(t)$, then we can proceed as follows:

```
> xsoln:=subs(soln,x(t));
```
$$xsoln := \mathbf{proc}(t)\ \ldots\ \mathbf{end}$$

```
> xsoln(1);
```
$$.8563831850074065$$

Similarly, we can force Maple to extract the approximate values for $y(t)$.

```
> ysoln:=subs(soln,y(t));
```
$$ysoln := \mathbf{proc}(t)\ \ldots\ \mathbf{end}$$

```
> ysoln(1);
```
$$1.571478273244474$$

If we want to see a list of approximate values then we can create a loop.

```
> for i from 0 by 0.2 to 1 do xsoln(i),ysoln(i) od;
```

1.5, 1.5

1.351719289423337, 1.538787611026623

1.210215295504144, 1.564878323230307

1.079246527364848, 1.578432790902777

.9609908986959353, 1.580233791564261

.8563065843455923, 1.571488776067847

It is also possible to plot the approximate solutions obtained from the **numeric** option of **dsolve**. We simply need to load the **plots** package and employ the **odeplot** command.

```
> with(plots):
> odeplot(soln,[[t,x(t)],[t,y(t)]],0..20,numpoints=100,view=[0..20,0..3]);
```

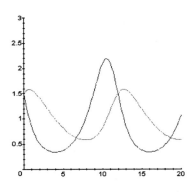

From the sketch above, it appears as though the period of the solution is roughly 13. The approximate solution curve $(x(t), y(t))$ is plotted below.

```
> odeplot(soln,[x(t),y(t)],0..13,numpoints=100,view=[0..3,0..3]);
```

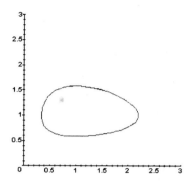

7.5 Summary

- The **dsolve** command can be used to find either explicit or numerical solutions of differential equations.

- The **odeplot** command can be used to plot the numerical solution produced by using the **numeric** option within **dsolve**.

- The **dfieldplot** and **phaseportrait** commands can be used to create direction fields and phase portraits.

- The **display** command can be used to simultaneously graph several plots.

7.6 Exercises

Use Maple to solve the following differential equations. If an initial value is given, plot the solution.

1. $\dfrac{dy}{dx} = \dfrac{xy + 3x}{x^2 + 1}$, $y(2) = 1$

2. $\dfrac{dy}{dx} = \dfrac{\ln(x)}{xy + xy^3}$

3. $y' + (\cos x)y = \cos x$

4. $y'' + \sin(y) = 0$, $y(0) = 1$, $y'(0) = 0$

5. Solve the equation $y' + y = \sin(x)$, with initial condition $y(0) = 0$, both numerically and exactly. Plot both solutions together using the range $-2 \le x \le 10$. How well does the numerical solution approximate the exact solution?

6. Plot the direction field of the differential equation $y' = x - y$. On a sheet of paper, hand sketch what you think the graph of the solution to this differential equation and the initial condition $y(0) = 0$ will look like. Then plot the solution and the direction field together.

7. Use Maple to sketch the direction field for the system

$$\begin{cases} x'(t) = -0.1x(t) - y(t) \\ y'(t) = x(t) - 0.1y(t) \end{cases}$$

Sketch the solution trajectories associated with a variety of initial data pairs. What do you notice about the behavior of the solution curves?

8. The purpose of this problem is to describe the phase portraits for the system

$$\begin{cases} x'(t) = ax(t) - by(t) \\ y'(t) = bx(t) + ay(t) \end{cases} \tag{7.2}$$

for a variety of choices of the parameters a and b.

(a) Start by randomly selecting a value for b. Then choose several negative values of a. For each of these values create a sketch of the associated phase portrait. What do these phase portraits have in common?

(b) Repeat this process with positive values of a.

(c) Repeat this process with $a = 0$.

(d) Test your conclusions in each of the parts above by selecting other values for b.

9. Show that if c_1 and c_2 are constants then

$$x(t) = c_1 e^{at} \cos(bt) - c_2 e^{at} \sin(bt)$$

$$y(t) = c1 e^{at} \sin(bt) + c_2 e^{at} \cos(bt)$$

solve the system (7.2). Does this reinforce your conclusions from exercise 8? Explain.

10. Consider the Lotka-Volterra system given by

$$\begin{cases} x'(t) = x(t)\,(1 - y(t)) \\ y'(t) = 0.3y(t)\,(x(t) - 1) \end{cases}$$

The phase portrait for this system was sketched earlier in this chapter. From looking at the phase portrait, it appears as though the positive solutions of this system are periodic. Use the **numeric** option of **dsolve** to determine the period of the solutions to this system corresponding to the initial data in the following table:

$x(0)$	0.2	0.4	0.6	0.8	1.2	1.4	1.6
$y(0)$	0.2	0.4	0.6	0.8	1.2	1.4	1.6

Chapter 8

Sequences and Series

Background Information: Read Chapter 12 in Stewart's **Calculus.**

In this section, we will learn how to use Maple's plot capabilities to explore the behavior of sequences. In addition, we will use Maple to sum various series. Two Maple commands used in this section are **seq** and **Sum**.

To construct a sequence of numbers whose individual terms are given by a formula, first input the formula that describes the terms as a function of n.

> `a:=n->1/n^2;`

$$a := n \rightarrow \frac{1}{n^2}$$

To have Maple construct the first five terms of this sequence, use the **seq** command. Its first argument is an expression for the terms of the sequence. Its second argument tells Maple which terms to construct.

> `seq(a(n),n=5..9);`

$$\frac{1}{25}, \frac{1}{36}, \frac{1}{49}, \frac{1}{64}, \frac{1}{81}$$

When the **seq** command is used, Maple increments the index variable—in this case n—through the integers 5-9. The value of the variable n at the end of the command is then 10. Make it a habit when using the **seq** command to always unassign the variable used. Thus, the following commands are preferred: **seq(a(n),n=5..9); n:='n':**. The index variable n is now free for use in the next sequence.

8.1   Limits of Sequences

Background Information: Read Section 12.1 in Stewart's **Calculus.**

119

One of the first problems we encounter is to determine the limiting value
of a sequence as $n \to \infty$. Maple is very good at calculating the limiting value
and displaying the sequence graphically. Let's examine the behavior of the
sequence $b_n = \dfrac{2n - 1}{3n + 6}$ by plotting the terms of this sequence as a function of
the parameter n.

> **b:=n->(2*n-1)/(3*n+6); disp:=seq([n,b(n)],n=1..50): n:='n':**

$$b := n \to \frac{2\,n - 1}{3\,n + 6}$$

These Maple commands construct the sequence of lists that are the Cartesian
coordinates of the points (n, b_n). Notice that the **seq** command was terminated
with a colon in order to suppress Maple's output. When first trying this to check
for typos, use a semicolon and have Maple display only the first few terms. Then
click back on the line and edit it to get all desired terms and use a colon instead.
To plot the points, enter the following command.

> **plot([disp],style=point);**

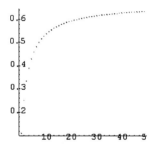

Maple's output is a plot of the first 50 terms of the sequence $\{b_n\}$. Notice that
we have enclosed the label **disp** in brackets. The argument [**disp**] is then an
ordered list of ordered pairs that Maple can plot. If you leave off the option
style=point, Maple will connect the points with straight lines. From the plot,
this sequence appears to have a limit as $n \to \infty$.

> **Limit(b(n),n=infinity); value(%);**

$$\lim_{n \to \infty} \frac{2\,n - 1}{3\,n + 6}$$

$$\frac{2}{3}$$

Recall that the **Limit** command (with a capital **L**) displays the limit and is used
in order to check for typos. The **value** command is then used to evaluate it.

This equation is so simple that Maple is not needed to compute the limit. However this same procedure can be used to handle more complicated problems.

8.2 Series

Background Information: Read Section 12.2 in Stewart's **Calculus**.

A second operation commonly performed on a sequence is to add some or all of its terms together. To sum the terms, use Maple's **Sum** command. If we let $a_n = 1/n^2$, as in the previous section, then

> **Sum(a(n),n=5..9); value(%);**

$$\sum_{n=5}^{9} \frac{1}{n^2}$$

$$\frac{737641}{6350400}$$

Let's try this command on the infinite series $\displaystyle\sum_{n=1}^{\infty} \frac{1}{n^2}$.

> **Sum(a(n),n=1..infinity); value(%);**

$$\sum_{n=1}^{\infty} \frac{1}{n^2}$$

$$\frac{1}{6}\pi^2$$

Maple may also be able to determine that a series diverges.

> **a:=n->1/n; Sum(a(n),n=1..infinity); value(%);**

$$a := n \to \frac{1}{n}$$

$$\sum_{n=1}^{\infty} \frac{1}{n}$$

$$\infty$$

Let's try one more.

> **a:=n->r^n; Sum(a(n),n=0..infinity); value(%);**

$$a := n \rightarrow r^n$$

$$\sum_{n=0}^{\infty} r^n$$

$$-\frac{1}{r-1}$$

Unfortunately, Maple did not warn us that this series converges to the above value only if $|r| < 1$. You might try to compute the value of the series $\sum_{n=0}^{\infty} 2^n$.

Even though Maple computes the above examples correctly, the sum of most infinite series cannot be computed exactly. For many series the goal is to discover whether or not the series converges and to compute approximate values for its sum when it does converge.

8.3   Convergence of Series

Background Information: Read Sections 12.3, 12.4 and 12.6 in Stewart's **Calculus.**

There are four very useful tests for the convergence of a series of positive terms: limit comparison, ratio, root, and integral.   In this section we will use Maple and these tests to determine whether or not a series of positive terms converges. Let us examine the series $\sum_{n=0}^{\infty} \dfrac{2^{3n}}{(2n+1)!}$.

> **a:=n->2^(3*n)/(2*n+1)!;**

$$a := n \rightarrow \frac{2^{(3n)}}{(2n+1)!}$$

Again, the terms are entered as functions of n. This allows us to refer back to individual terms when we use the ratio or root test.

> **Limit(a(n+1)/a(n),n=infinity); value(%);**

$$\lim_{n \to \infty} \frac{2^{(3n+3)}(2n+1)!}{(2n+3)!\,2^{(3n)}}$$

$$0$$

Since the limit of the ratio of consecutive terms is less than one, the series converges. The same result is obtained when the limit of the nth root of the terms is computed.

```
> Limit(a(n)^(1/n),n=infinity); value(%);
```

$$\lim_{n \to \infty} \left(\frac{2^{(3n)}}{(2n+1)!} \right)^{\left(\frac{1}{n} \right)}$$

$$0$$

When the limit in the ratio or root test is different from 1, we know whether the series converges or diverges. When this limit is 1, another test is needed. For example, in the series $\displaystyle\sum_{n=2}^{\infty} \frac{\ln(n)}{n^2}$ the limit of the ratios and the limit of the nth roots is 1. Instead, since $\ln(n)$ grows slower than any positive power of n, we can try a comparison test with the series $\sum_{n=2}^{\infty} \frac{1}{n^{3/2}}$.

```
> a:=n->ln(n)/n^2; b:=n->1/n^(3/2);
> Limit(a(n)/b(n),n=infinity); value(%);
```

$$a := n \to \frac{\ln(n)}{n^2}$$

$$b := n \to \frac{1}{n^{3/2}}$$

$$\lim_{n \to \infty} \frac{\ln(n)}{\sqrt{n}}$$

$$0$$

Since this p-series converges $(3/2 > 1)$ and the limit of the ratios of the first series to the p-series is finite, we know the series $\displaystyle\sum_{n=2}^{\infty} \frac{\ln(n)}{n^2}$ converges. Alternatively, since the terms of this series are positive and decreasing, we can try the integral test on this series.

```
> Int(a(n),n=1..infinity); value(%);
```

$$\int_1^{\infty} \frac{\ln(n)}{n^2} \, dn$$

$$1$$

Since the integral is finite, the series $\displaystyle\sum_{n=2}^{\infty} \frac{\ln(n)}{n^2}$ converges.

8.4   Error Estimates

Background Information: Read Section 12.3 in Stewart's **Calculus**.

One of the more useful aspects of the integral test is its ability to estimate the error involved in using a partial sum to approximate a series. For example, we saw earlier that the sum of the series $\sum_{n=1}^{\infty} \frac{1}{n^2}$ is $\frac{\pi^2}{6}$. Suppose we didn't know this and added up the first 50 terms of the series. How close are we to the actual sum?

```
> a:=n->1/n^2; Sum(a(n),n=1..50); S50:=evalf(%);
> error:=evalf(Pi^2/6-S50);
```

$$a := n \rightarrow \frac{1}{n^2}$$

$$\sum_{n=1}^{50} \frac{1}{n^2}$$

$$S50 := 1.625132734$$

$$error := .019801334$$

Using the ideas underlying the integral test

$$\int_{51}^{\infty} \frac{dx}{x^2} \leq \sum_{n=51}^{\infty} \frac{1}{n^2} \leq \int_{50}^{\infty} \frac{dx}{x^2}$$

The sum is an upper Riemann sum for $\int_{51}^{\infty} \frac{1}{x^2} dx$ and a lower Riemann sum for $\int_{50}^{\infty} \frac{1}{x^2} dx$ as illustrated below.

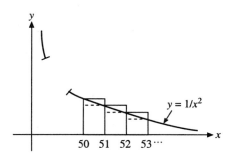

> Int(1/x^2,x=51..infinity); l:=evalf(%);

$$\int_{51}^{\infty} \frac{1}{x^2}\, dx$$

$$l := .01960784314$$

> Int(1/x^2,x=50..infinity); u:=evalf(%);

$$\int_{50}^{\infty} \frac{1}{x^2}\, dx$$

$$u := .02000000000$$

Thus we have

$$lower = l + S50 \leq \sum_{n=1}^{\infty} \frac{1}{n^2} \leq u + S50 = upper$$

We can use the midpoint of the interval [*lower, upper*] as an approximate value for the sum of the series. The error of this approximation is at most (*upper* − *lower*)/2, which is half the width of this interval.

> lower:=l+S50; upper:=u+S50;
> midpoint:=(lower+upper)/2;
> maxerror:=(upper-lower)/2;

$$lower := 1.644740577$$

$$upper := 1.645132734$$

$$midpoint := 1.644936656$$

$$maxerror := .0001960785$$

From above we see that $\pi^2/6 \approx 1.644936656$.

Note. This averaging method gives us a much better approximation of the infinite series than that obtained by just using the partial sum of the first 50 terms, as we did earlier. Moreover, any extra work involved is minimal.

8.5 Taylor Polynomials

Background Information: Read Sections 12.10 and 12.12 in Stewart's **Calculus**.

An extremely useful idea in mathematics is the approximation of complicated functions with simpler ones. In this section the simpler functions are
polynomials.

Suppose we want the fifth degree Taylor polynomial of $\sin(x)$. The following
Maple commands will construct this polynomial.

> **taylor(sin(x),x=0,6); p:=convert(%,polynom);**

$$x - \frac{1}{6}x^3 + \frac{1}{120}x^5 + O(x^6)$$

$$p := x - \frac{1}{6}x^3 + \frac{1}{120}x^5$$

A few words about the syntax are in order. Notice that the Maple command
taylor has three parameters: the expression whose Taylor polynomial we want;
the point that we expand about; and an integer. This last parameter—in this
case 6—is one more than the degree of the desired Taylor polynomial. In Maple
this number refers to the order of the error of the approximating polynomial.
Notice too that the Maple command **convert** enables us to assign the polynomial
to a variable without having to retype it.

Taylor polynomials are particularly useful because they not only give an
effective way to approximate a function, but also the form of the error term
can often lead to valuable estimates as to how well we have approximated our
function. For example, suppose we want to approximate the sine function on the
interval $[0, \pi]$ with its Taylor polynomial of degree 7, and we wish to know how
good an approximation we have. First we decide to expand about the midpoint
of the interval.

> **taylor(sin(x),x=Pi/2,8); p:=convert(%,polynom);**

$$1 - \frac{1}{2}\left(x - \frac{1}{2}\pi\right)^2 + \frac{1}{24}\left(x - \frac{1}{2}\pi\right)^4 - \frac{1}{720}\left(x - \frac{1}{2}\pi\right)^6 + O\left(\left(x - \frac{1}{2}\pi\right)^8\right)$$

$$p := 1 - \frac{1}{2}\left(x - \frac{1}{2}\pi\right)^2 + \frac{1}{24}\left(x - \frac{1}{2}\pi\right)^4 - \frac{1}{720}\left(x - \frac{1}{2}\pi\right)^6$$

The formula for the remainder, $\dfrac{f^{(n+1)}(\xi)}{(n+1)!}(x - \pi/2)^{(n+1)}$, involves in this case
the eighth derivative of the sine function, evaluated at some point ξ between x
and $\pi/2$.

> **diff(sin(x),x$8); subs(x=xi,%);**

$$\sin(x)$$

$$\sin(\xi)$$

We know the absolute value of sine is never larger than 1. Moreover, the distance between x and $\pi/2$ is always less than or equal to $\pi/2$ (recall that x lies in the interval $[0, \pi]$). Since we are expanding about the point $\pi/2$, we can therefore estimate the error as follows.

> **evalf((Pi/2)^8/8!);**

$$.0009192602758$$

Thus our seventh degree polynomial is uniformly within 0.00092 of the value of the sine function on the interval $[0, \pi]$. To visualize this, we plot the sine function together with the Taylor polynomial.

> **plot({sin(x),p},x=-Pi/2..3*Pi/2);**

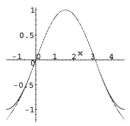

8.6 Summary

The new Maple commands in this chapter are **seq**, **Sum**, **taylor** and **convert**. You should know the syntax of these commands, know how to plot sequences, and know the new use of the **convert** command for dealing with Taylor polynomials.

8.7 Exercises

For Exercises 1–5, plot each of the sequences. Try to determine whether or not the sequence has a limit as $n \to \infty$ and what the limiting value is. Then have Maple compute each limit that exists.

1. $a_n = \dfrac{2 + (-1)^n n^2}{n^2 - 3n + 4}$

2. $a_n = \dfrac{\ln (n)}{n^{1/3}}$

3. $a_n = \dfrac{n!}{200^n}$

4. $a_n = \dfrac{2n^3 - 6n^2 + 15}{n^4 + 18n^3 - 6}$

5. $a_n = \left(1 + \dfrac{2}{n}\right)^{1/n}$

In Exercises 6–10 sum the given series using Maple. Then, simultaneously plot the sequence of terms and the sequence of partial sums for each of the series.

6. $\displaystyle\sum_{n=1}^{\infty} \dfrac{1}{3^n}$

7. $\displaystyle\sum_{n=1}^{\infty} \dfrac{(-1)^{n-1}}{n}$

8. $\displaystyle\sum_{n=1}^{\infty} \dfrac{1}{n^k}$, for various values of k

9. $\displaystyle\sum_{n=1}^{\infty} \dfrac{n^3}{2^n}$

10. $\displaystyle\sum_{n=1}^{\infty} \dfrac{1}{1 + n^2}$

In Exercises 11–16, decide whether the given series converges. In Exercises 13–16, give the values of x for which the series converges. For each of the series below, try as many of the convergence tests discussed in this chapter as seem applicable.

11. $\displaystyle\sum_{n=1}^{\infty} \dfrac{n^2}{\sqrt{n^5 + n^2 + 2}}$

12. $\displaystyle\sum_{n=1}^{\infty} \dfrac{n}{\ln (n^n)}$

13. $\displaystyle\sum_{n=1}^{\infty} (\sqrt[n]{2} - 1)(x - 2)^n$

14. $\displaystyle\sum_{n=1}^{\infty} \frac{(-1)^n x^{2n}}{n \cos(n)}$

15. $\displaystyle\sum_{n=1}^{\infty} 3^n x^n$

16. $\displaystyle\sum_{n=1}^{\infty} \frac{(-1)^n x^n}{n 2^n}$

For Exercises 17–19, find the requested Taylor polynomial and estimate the error in approximating the given function with the Taylor polynomial on the given interval. Then plot both the function and the Taylor polynomials on the same coordinate axes.

17. The fourth degree Taylor polynomial of $\cos(3x)$ about $x = \pi/4$ on the interval $[0, \pi]$.

18. The first, fifth, and fifteenth degree Taylor polynomials of $\dfrac{x^4 - 16x^2 + 2x - 5}{x^2 - 6}$ about $x = 1$. Plot all of them and the given function on the same graph for the interval $[-1, 2]$.

19. The 100th degree Taylor polynomial of $x^3 - 2x^2 + 15x - 6$ about $x = 0$ and also about $x = 5$, on the interval $[-2, 8]$.

20. Estimate the value of the series $\displaystyle\sum_{n=1}^{\infty} \frac{1}{n^3}$ by using the sum of the first 25 terms of the series and the averaging method discussed in section 8.4.

21. Expand the function $\sin(x)$ about $x = \pi/2$ in a Taylor polynomial of sixth degree and also of seventh degree. The seventh degree polynomial should be more accurate than the sixth. Is it? How much extra work is involved in evaluating this seventh degree Taylor polynomial than in evaluating the sixth degree one?

22. Compute the Taylor polynomial of degree 5 centered at $a = 0$ for the function $f(x) = \sin(x)$. Name this polynomial $p(x)$. Now evaluate $p(x^2)$ and compare it to the 10th degree Taylor polynomial centered at $a = 0$ for the function $f(x) = \sin(x^2)$. What do you observe?

23. Compute the Taylor polynomial of degree 6 centered at $a = 0$ for the function $f(x) = \cos(x)$. Name this polynomial $p(x)$. Now evaluate $p(x^3)$ and compare it to the 18th degree Taylor polynomial centered at $a = 0$ for the function $f(x) = \cos(x^3)$. What do you observe?

24. Compute the Taylor polynomial of degree 7 centered at $a = 0$ for the function $f(x) = e^x$. Name this polynomial $p(x)$. Now evaluate $p(x^4)$ and compare it to the 28th degree Taylor polynomial centered at $a = 0$ for the function $f(x^4)$. What do you observe?

25. Use the previous 3 problems to make a conjecture concerning the relationship of the nth degree Taylor polynomial centered at $a = 0$ for a function $f(x)$, and the $(k*n)^{th}$ degree Taylor polynomial centered at $a = 0$ for the function $f(x^k)$.

26. Use a an appropriate degree Taylor polynomial centered at $a = 0$ for the function $f(x) = \sin(x^2)$ to approximate

$$\int_{-1}^{1} \sin(x^2)\,dx$$

to 7 decimal places of accuracy.

27. Repeat the previous problem for the integral

$$\int_{-\pi}^{\pi} e^{-x^2}\,dx$$

28. It is not always clear how to choose the value of the centering point for a Taylor polynomial. Consider the function $f(x) = e^{3x} + 2\sin(x)$ on the interval $[-1, 1]$. Let $p_{3,a}(x)$ denote the generic 3rd degree Taylor polynomial centered at $x = a$ for $f(x)$. Find the value of a so that

$$\int_{-1}^{1} [f(x) - p_{3,a}(x)]^2\,dx$$

is minimized.

Chapter 9

Polar Plots

Background Information: See Chapter 11 in Stewart's **Calculus.**

This chapter begins with a review of the Maple commands that are used to plot parametrized curves, and it concludes with a brief discussion of polar plots.

9.1   Parameterized Curves and Polar Plots

Some figures, such as circles and ellipses, are not the graphs of functions. Instead, these figures are more conveniently described by *parametric equations*, which are of the form

$$x = f(t) \quad y = g(t), \quad a \le t \le b$$

where f and g are functions of the parameter t, and a and b are numbers. As a simple example, a circle of radius 3 centered at the origin is described by the parametric equations

$$x = 3\cos(t) \quad y = 3\sin(t), \quad 0 \le t \le 2\pi.$$

Parametric equations are often used to describe the position of a particle at time t.

As stated in chapter 1, to plot the parametric equations

$$x = f(t) \quad y = g(t), \quad a \le t \le b$$

first define the functions f and g (using the formula **f := t ->**, as in Section 2.2), and then type the command **plot([f(t),g(t),t=a..b]);** Note that square brackets [ ] are used for a parameterized plot (the curly braces { } are used to plot the graphs of two functions $y = f(x)$ and $y = g(x)$). If f and g are expressions in t rather than functions of t, the syntax is the same, except that f and g would be entered rather than $f(t)$ and $g(t)$.

Example 1. To plot a circle of radius 3 centered at the origin, type

> **plot([3*cos(t),3*sin(t),t=0..2*Pi]);**

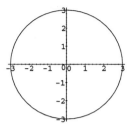

Your circle may look deformed (more like an ellipse). This is due to the fact that the scales for the x- and y-axes are different. To correct this problem, use your mouse to click on the **Projection** menu at the top of your plot, and then click on the word **Constrained**.

Example 2. Plot the polar equation

$$r = \cos(3\theta), \quad 0 \le \theta \le 2\pi$$

Solution. Recall that a polar curve is an example of a parametric curve with the parameter θ playing the role of t. Since it is easier to type t rather than θ, we will use t for the parameter. To plot the polar equation $r = \cos(3t)$, convert from polar coordinates to rectangular coordinates using

$$x = r \cos(t) \qquad y = r \sin(t)$$

and enter the commands

> **r:=cos(3*t): plot([r*cos(t),r*sin(t),t=0..2*Pi],scaling=constrained);**

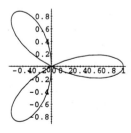

Example 3. Plot the parabola given in polar form by $r = 1/(1 - \sin(t))$.
Solution. First enter the expression for r and then enter the rectangular coordinate expressions for this parabola.

> r:=1/(1-sin(t)); x:=r*cos(t); y:=r*sin(t);

$$r := \frac{1}{1 - \sin(t)}$$

$$x := \frac{\cos(t)}{1 - \sin(t)}$$

$$y := \frac{\sin(t)}{1 - \sin(t)}$$

Now enter the command **plot([x,y,t=0..2*Pi]);**. Note that the scale of the plot is large due to the fact that the expression r is large for t near $\pi/2$. To view a portion of the graph near the origin, enter the following modification of the **plot** command.

> plot([x,y,t=0..2*Pi],-5..5,-5..5);

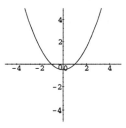

A command, called **polarplot**, is specialized for polar coordinate plots. To execute this, you must first load the plot package (i.e., enter **with(plots);**). Then, for example, the above parabola can be plotted by typing **polarplot(r,t=0..2*Pi);**. To specify the ranges $-5 \le x \le 5$ and $-5 \le y \le 5$, the view option must be used by entering

> polarplot(r,t=0..2*Pi,view=[-5..5,-5..5]);

9.2   Exercises

1. Plot the ellipse $x = 3\cos(t)$ and $y = \sin(t)$ using constrained scaling.

2. Plot a piece of the hyperbola parameterized by $x = \cosh(t)$ and $y = \sinh(t)$, for $-2 \le t \le 2$. Recall that $\cosh(t) = (e^t + e^{-t})/2$ and $\sinh(t) = (e^t - e^{-t})/2$.

3. Graph each of the following equations by hand and then check your answers with a Maple plot. If the graph is a conic section (ellipse, hyperbola,

or parabola), then give the location of the foci or focus. You may wish to use **style=point** on (b)–(e).

(a) $x^2 + 2y^2 = 4$

(b) $r = \dfrac{3}{1 - \cos(t)}$

(c) $r = \dfrac{2}{1 - 2\sin(t)}$

(d) $r = \dfrac{4}{1 + 2\cos(t)}$

(e) $r = \dfrac{4}{1 - \sin(t + \pi/3)}$

(f) $r = \dfrac{2}{3 + \cos(t + \pi/4)}$

(g) $r = 2 + \cos(t)$

4. Recall that the formula for computing the area of the region that lies between a polar graph $r = r(t)$ and the origin and between the angles $t = a$ and $t = b$ is given by

$$\frac{1}{2} \int_a^b r(t)^2 \, dt$$

Set up and evaluate the integral(s) required to find the area that is inside both the parabola $r1(t) = \dfrac{3}{1 - \cos(t)}$ and the circle $r2(t) = 8$. *Hint:* First plot these two graphs to estimate the angles where the two curves cross. Next use **fsolve** to find the angles. Then perform the appropriate integrals using Maple's **Int** and **evalf** commands.

5. Recall that the arc length of a polar graph $r = r(t)$, $a \le t \le b$, is given by

$$\int_a^b \sqrt{r(t)^2 + r'(t)^2} \, dt$$

The formulas for finding the surface area of revolution of this polar graph about the x- and y-axes are, respectively

$$2\pi \int_a^b y(t) \sqrt{r(t)^2 + r'(t)^2} \, dt$$

$$2\pi \int_a^b x(t) \sqrt{r(t)^2 + r'(t)^2} \, dt$$

provided a and b are appropriate (see Exercise 6). Here, $x(t) = r(t)\cos(t)$ and $y(t) = r(t)\sin(t)$.

(a) Set up and evaluate the integral involved in calculating the arc length of the curve $r = 2 + \cos(t)$, $0 \le t \le \pi/2$.

(b) Set up and evaluate the integral involved in calculating the area of the surface obtained by revolving the curve $r = 2 + \cos(t)$, $0 \le t \le \pi/2$, about the x-axis.

(c) Set up and evaluate the integral involved in calculating the area of the surface obtained by revolving the curve $r = 2 + \cos(t)$, $0 \le t \le \pi/2$, about the y-axis.

6. Set up and evaluate the integral involved in calculating the area of the surface obtained by revolving the petal of the rose $r = \cos(3t)$, $-\pi/6 \le t \le \pi/6$, about the x-axis. Be careful in choosing your limits of integration.

Chapter 10

Programming with Maple

We have indicated in earlier chapters that Maple has a built-in programming language. In fact, it is possible to use Maple to create highly structured programs that can perform a large number of useful tasks. Most importantly, any program written in Maple can call any of the high level packages available in Maple, or any program that you have written in Maple and included in the same session.

This short chapter is intended as an introduction to the Maple programming environment. We do not intend this chapter to be a tutorial on structured programming. Rather, our intent is simply to familiarize the reader with some of the programming capability available within Maple. As a result, we will assume the reader has some previous experience with structured programming.

10.1 Introduction to Maple Procedures

Maple programs are called procedures, and they all have the structure

proc(arguments)/statements/end;

The statements that are included in procedures can be any sequence of valid Maple commands, with each executable statement (except optionally the last one) ending in either a colon or a semicolon. As we will see below, these commands include not only those with which you have become familiar in earlier sections, but also standard commands that would be expected in any highly structured programming language.

You certainly realize that you have been writing simple Maple procedures in earlier chapters, but you might be puzzled because they do not seem to have the form given above. For example, you have probably defined a function such as $f(x) = x^2$ using the command

```
> f:=x->x^2;
```

$$f := x \rightarrow x^2$$

When you write a procedure in this manner, you are actually using Maple shorthand to write a procedure in **proc(arguments)/statements/end;** format. The arrow symbol tells Maple to create a procedure that has the single argument (or input) x and gives the output x^2. We could use the structure listed for Maple procedures above to write this procedure as

```
> f:=proc(x) x^2 end;
```

f := proc(x) x^2 end

This gives the same result as the arrow notation above. To see this, you can simply enter

```
> f(x);
```

$$x^2$$

Now let's create some procedures that are slightly more complex than the simple example given above. The first of these illustrates a simple use of the **for/do** looping structure within Maple. This structure has the form

| for <name> || from <expr> || by < expr> || to < expr > |
| while <expr> | do <statement sequence> od;

where | | indicates optional phrases.

Example 1. Suppose you need a simple procedure that can be used to determine whether a pattern emerges when higher order derivatives of a function are computed. We give an example procedure that can be used to list the first k derivatives of a given function f.

```
> der_pat:=proc(f,k)
>     local i, x;
>         for i from 1 to k do
>             print((D@@i)(f)(x));
>         od;
>     end;
```

der_pat := proc(f,k) local i,x; for i to k do print((D @@ i)(f)(x)) od end

Note that we used local variables in the creation of our procedure. Let's use this program to look for a possible pattern in the derivatives of $e^x \sin(x)$. To this end, we will ask for the first four derivatives of this function.

```
> f:=proc(x) exp(x)*sin(x) end:
> der_pat(f,4);
```

$$e^x \sin(x) + e^x \cos(x)$$

$$2\,e^x \cos(x)$$

$$2\,e^x \cos(x) - 2\,e^x \sin(x)$$

$$-4\,e^x \sin(x)$$

We observe that the fourth derivative of the function is -4 times the function. Similarly, we see that 2 times the first derivative of the function minus the second derivative of the function is 2 times the function. You might try using this information to write a variety of differential equations that the function $f(x) = e^x \sin(x)$ satisfies.

Example 2. Suppose you need a procedure that will plot a function along with its tangent lines at several points. We create such a procedure below; its inputs are the function, the interval for the plot, and a numeric array of x coordinates for the points. Note that we have included a colon after the **end** statement to suppress Maple's echo.

```
> tan_plot:=proc(f,a,b,x_array)
>     local plist,i,x,xp,n;
>         xp:=x_array; #xp used for readability
>         n:=nops(x_array);
>         plist:=f(x);
>         for i from 1 to n do
>             plist:=plist,D(f)(xp[i])*(x-xp[i])+f(xp[i]);
>         od;
>         plot({plist},x=a..b);
>     end:
```

We can use this program to plot $4\sin(x) + \cos(2x)$ along with its tangent lines at $x = -1, 2, 4$ on the interval $[-3, 6]$ as follows.

```
> f:=x-> 4*sin(x)+cos(2*x):
> tan_plot(f,-3,6,[-1,2,4]);
```

Again note that we employed local variables in our procedure. Also, we may have used a command that you are not familiar with, namely, **nops**. This command returns the number of entries in **x_array**. Finally, note that, although there are several executable statements with semicolons, we only see the output from the last one. That is because a Maple procedure only returns the value of the last executable statement.

We close this section with a simple illustration that built-in Maple procedures can be used to simplify programming. First, a procedure is written to accomplish the required task, and then a slightly advanced use of built-in Maple procedures is employed to obtain the same result. We also use this example to demonstrate how more than one output can be returned from a procedure without using a print statement. This is important if you want to assign the output to a variable and perform Maple operations on it.

Example 3. Assume you need a procedure that evaluates a function at a variety of numbers and then sums the function values. To create such a procedure, we need to pass the function and an array of the numbers to a program that performs this operation. One such (crude) procedure can be written as follows.

```
> sum_func:=proc(g,num_array)
>     local n,i,s,a;
>         n:=nops(num_array);
>         a:=num_array;
>         s:=0;
>         for i from 1 to n do
>                 s:=s+g(a[i]);
>         od;
>         s;
>     end:
```

Notice that we used a colon following the end command to suppress Maple's echo.

If we want to evaluate the function $f(x) = x^2 + x$ at the numeric values $2, 3, 7, 9, 13, 17, 20$ and sum the resulting numbers, we can do the following.

```
> f:=proc(x) x^2+x end:
> sum_func(f,[2,3,7,9,13,17,20]);
                        1072
```

The following procedure is more efficient and takes advantage of some built-in high level commands in Maple. If you are not familiar with the **seq** and **sum** commands, this might be a good opportunity to do some investigation.

```
> sum_func1:=proc(g,num_array)
>     local n,k,a,i;
>         n:=nops(num_array);
>         a:=[seq(g(i),i=num_array)];
>         sum(a[k],k=1..n);
>     end:
```

We can verify that this procedure produces the same result as follows.

```
> sum_func1(f,[2,3,7,9,13,17,20]);
                        1072
```

It worked!

We also mentioned above that Maple only returns the result of the last executable statement in a procedure. If you want to see more of the output from the procedure, then you might try implementing the procedure given below.

```
> sum_func2:=proc(g,num_array)
>     local n,k,a,i;
>         n:=nops(num_array);
>         a:=[seq(g(i),i=num_array)];
>         RETURN(n,a,sum(a[k],k=1..n));
>     end:
```

You should try this version to see the difference in the output. As mentioned above, the advantage of using the **RETURN** statement, as opposed to **print**, is that the output from **RETURN** can be assigned to a variable. See the Help facility for other uses of the **RETURN** statement.

There are certainly a large number of Maple programming structures that we have not included in this short introduction. For example, Maple has built-in **if/then/else** capability. The syntax for the **if/then/else** structure is given by

if < conditional expression> then < statement sequence>
| elif < conditional expression> then < statement sequence> |
| else < statement sequence > | fi;

Again note that | | indicates an optional phrase.

An example that employs this structure and others is given in the next section. We encourage you to explore all of Maple's programming capability by clicking on the **Help Browser** menu and selecting the topic **Programming....**

10.2 Creating Interactive Programs in Maple

The examples in this section indicate how interactive procedures can be written in Maple. The primary programming structure employed is the **readstat** command. We comment here (as well as in the programs below) that responses to prompts created by **readstat** must use correct Maple syntax, including the use of a colon or semicolon. In addition, if the response is text, then it must be included in backquotes.

Example 1. To illustrate the use of the **readstat** command, we begin with something very simple. We doubt that you will actually use this procedure in your upper level science/engineering courses, but it illustrates the application of some basic Maple programming constructs. Namely, this procedure introduces the **rand** command (for generating random numbers), demonstrates the use of **if/then/else** and logical **and/or** structure, and illustrates that it is possible to write procedures that do not require that arguments be passed to them. The following procedure asks a simple question and then checks the response of the user. The process continues until three correct responses are given. You should note the difference in the output from the commands **lprint** and **print**.

```
> simple_1:=proc()
>     local ran,a,num_correct,ans;
>         num_correct:=0;
>         lprint('Respond to each question below with either
>              "yes" or "no"');
>         lprint('Be sure to follow your response with a semi-colon.');
>         ran:=rand(0..20)-10:
>         while num_correct<3 do
>             a:=ran();
>             lprint('Consider the number');
>             print(a);
>             ans:=readstat('Is this number less than zero?');
>             if  a<0 and ans='yes' or a>=0 and ans='no' then
>  num_correct:=num_correct+1;
>                 lprint('You have answered', num_correct, 'question(s)
> correctly');
>                 else lprint('Your answer is incorrect');
>             fi;
>         od;
>         lprint('Congratulations!  You are done.');
>     end:
```

You can run the program above by typing **simple_1();**.

One problem with creating interactive programs in Maple is that the response must employ correct Maple syntax. To avoid problems, it is possible to use error trapping. We refer the interested reader to the Maple command **traperror**.

We close this section with a somewhat more interesting program that employs interaction.

Example 2. The procedure shown below (and continued at the top of the next page) can be used to interactively create an animation of the convergence of Taylor polynomial approximations to a function. Notice that this procedure loads the **plots** package. Again, no arguments are passed to the procedure.

```
> taylor_converge:=proc()
>      local f_expr,a,n_array,n,i,tp,plist,xinterv,yinterv;
>          with(plots):
>          lprint(' ');
>          lprint('This program can be used to create an animation
> of the convergence');
>          lprint('of Taylor polynomial approximations to a function.
> Make sure you end');
>          lprint('your responses with a ";" . ');
>          lprint(' ');
>          lprint('Input your function in expression form. ');
```

```
>          f_expr:=readstat('For example, you might input  sin(x); . ');
>          a:=readstat('Where do you want your Taylor polynomials
> centered? ');
>          lprint(' ');
>          lprint('Give an array of polynomial degrees.');
>          n_array:=readstat('For example, you might give [1,5,7,15]; . ');
>          xinterv:=readstat('Give the x plot interval in the form a..b; . ');
>          yinterv:=readstat('Give the y plot interval in the form a..b; . ');
>          lprint(' ');
>          lprint('A plot window will appear in a moment.');
>          lprint('Click on the arrow button directly below the edit menu
> to play the animation');
>          n:=nops(n_array);
>          plist:=NULL;
>          for i from 1 to n do
>              tp:=convert(taylor(f_expr,x=a,n_array[i]+1),polynom);
>              plist:=plist,plot({f_expr,tp},x=xinterv,y=yinterv):
>          od;
>          display([plist],insequence=true);
>      end:
```

A sample interaction with this program is given below (only user input is given).

```
> taylor_converge();
> sin(x)+cos(2*x);
> 1;
> [1,5,10,15,20,25];
> -6..6;
> -6..6;
```

Of course the length of the program above can be shortened dramatically by removing the interaction. In noninteractive form it appears as

```
> tay_appr:=proc(f_expr,a,n_array,xinterv,yinterv)
>      local n,plist,i,tp;
>          with(plots):
>          n:=nops(n_array);
>          plist:=NULL;
>          for i from 1 to n do
>               tp:=convert(taylor(f_expr,x=a,n_array[i]+1),polynom);
>               plist:=plist,plot({f_expr,tp},x=xinterv,y=yinterv):
>          od;
>          display([plist],insequence=true);
>      end:
```

You might want to try the command

```
> tay_appr(sin(x),0,[1,5,11,19,27],-10..10,-3..3);
```

10.3   Concluding Remarks

We have given only a brief glimpse of the programming capability of Maple. You should explore Maple's ability to read data from files, to run and manipulate programs outside of Maple, and to export portions of your Maple sessions in a variety of very useful formats.

Finally, you might have one important question that we have not addressed. Namely, if you create a program in Maple then you might naturally ask how you can use it in a different Maple session. The answer is very simple. Take the Maple programs that you create and place them in one or several files. For example, you might place the interactive Maple procedure **taylor_converge()** along with any other procedures related to Taylor polynomials in a file named taylor.txt. When you want to use these programs, simply invoke the Maple command **read('taylor.txt');**. Every Maple procedure in that file will be read into your Maple session and will be available for your use.

Chapter 11

Troubleshooting Tips

The Top Ten Headaches Reported by Beginning Maple Users:

1. Missing or incorrect punctuation.

2. Confusion between functions and expressions.

3. Re-executing statements to avoid retyping them.

4. Forgetting to save a Maple session before it's too late, and losing it.

5. Trouble using on-line help.

6. Forgetting to include library packages.

7. Confusion between exact and approximate calculations.

8. Trying to get Maple to do too much.

9. Making mistakes when entering a procedure.

10. Forgetting **od** after **do** and **fi** after **if**.

This list is based on the actual experiences of many new Maple users, mostly freshman and sophomore mathematics students and their instructors. They brought with them a wide range of computer experience. However, Maple has its own idiosyncracies, and certain problems seemed to arise, repeatedly, for just about everybody. It is hoped that by having these common errors pointed out to you, along with examples of how they occur, how to spot them, and tips for avoiding them, you will more quickly and painlessly achieve proficiency with this powerful software tool.

Before considering specific examples of common errors, we mention that many times such errors can be prevented by simply changing the font size at the beginning of a Maple session. This sounds trivial, but it's amazing how many students don't think about it, and suffer eyestrain and wasted time correcting errors caused by reading difficulties. In the same vein, some have found it helpful to generously insert blank spaces in lines of Maple statements to improve readability.

11.1   Missing or Incorrect Punctuation

11.1.1   Missing Semicolon or Colon at the End of a Line

The following is a very common error.

```
> f := x^2 + 1
> g := x^3 + 2;

syntax error:
g := x^3 + 2;
^
```

At first you stare at the second line, trying to spot the error, but you can't, because there's nothing wrong with it. Eventually, you spot the missing ; on the line above it. Just add the semicolon and re-execute the two lines.

Here is a similar example, which we will assume you have typed and entered (by pressing the ⟨ENTER⟩ key).

```
> sqrt(Pi + 5)
```

You notice the missing semicolon, so you put one in and hit the ⟨ENTER⟩ key again.

```
> sqrt(Pi + 5);

syntax error:
sqrt(Pi + 5);
^
```

Again, you see nothing wrong with the line. The problem is that Maple responded somewhat as if you had entered the following statement, since it treated the second entry as a continuation of the first.

```
> sqrt(Pi + 5)  sqrt(Pi + 5);

syntax error:
sqrt(Pi + 5)  sqrt(Pi + 5);
                 ^
```

The remedy is to simply re-enter the line.

```
> sqrt(Pi + 5);
```
$$\sqrt{\pi + 5}$$

Tip: Try to get into the habit of pausing to examine the output from each line you enter, and asking yourself if it is reasonably close to what you were expecting. Unfortunately, this doesn't always work, since sometimes there isn't supposed to be any output—for example, as when you're entering a *procedure* (see Section **11.9**).

11.1.2 Missing Parentheses

Maple makes unmatched parentheses fairly easy to find, as the following example shows.

```
> x*sqrt(2*x + 1) / ( (x^2 + 5)*(x + 1);

syntax error:
x*sqrt(2*x + 1) / ( (x^2 + 5)*(x + 1);
                                     ^
```

The following example shows how parentheses are used to ensure that certain calculations are carried out correctly. Suppose you want Maple to calculate $\sum_{k=1}^{10} \frac{1}{k(k+2)}$, and you enter

```
> sum(1/k*(k+2), k=1..10);
```
$$\frac{19981}{1260}$$

This is not correct, and you might have caught the error by noticing that the output is too large a number. Another way is to use the following *inert* form of **sum**.

```
> Sum(1/k*(k+2), k=1..10);
```
$$\sum_{k=1}^{10} \frac{k+2}{k}$$

This makes it easy to spot the missing parenthesis, so that you can enter the corrected statement.

```
> Sum(1/(k*(k+2) ), k=1..10);
```
$$\sum_{k=1}^{10} \frac{1}{k(k+2)}$$

```
> value(%);
```
$$\frac{175}{264}$$

11.1.3   Missing * for Multiplication

Forgetting a *** for multiplication** can lead to a variety of errors. For example

```
> y := 2(x+1) + x^2;
```
$$y := 2 + x^2$$

```
> y := (x+1)(x+2);
```
$$y := \mathrm{x}(x+2) + 1$$

```
> y := (2+x)(3+5);
```
$$y := \mathrm{x}(8) + 2$$

```
> y := (2+3)(3+5);
```
$$y := 5$$

In the first and last examples, the second factor is apparently just ignored. In the second and third examples, the second factor is treated as an argument of a function named x (notice the different fonts for the x's).

11.2 Confusion between Functions and Expressions

We will illustrate some of the errors that are typically made. In the following, **f** is the label of an expression, while **g** and **h** are labels (or names) for functions.

```
> f := x^2 + 1;
```

$$f := x^2 + 1$$

```
> g := x -> x^2 + 1;
```

$$g := x \to x^2 + 1$$

```
> h := x -> if x<1 then x^2+1 else 3*x fi;
 h := proc(x) options operator,arrow; if x  < 1 then x^2+1
else 3*x fi end
```

Variables in expressions are given numerical values with the **subs** command, while functions are evaluated using standard function notation.

```
> subs(x=2,f);
```
$$5$$

```
> g(2);
```
$$5$$

The following illustrates the errors that occur when the syntax appropriate for a function is used with an expression.

```
> f(x);
```
$$x(x)^2 + 1$$

```
> f(1);
```
$$x(1)^2 + 1$$

It is easy to convert expressions into functions, and vice versa.

```
> f1 := unapply(f,x);
```

$$f1 := x \rightarrow x^2 + 1$$

```
> g1 := g(x);
```

$$g1 := x^2 + 1$$

However, an error results when you try to change **h** into an expression the same way, by evaluating it at **x**.

```
> h1 := h(x);
```

```
Error, (in h) cannot evaluate boolean
```

This is because, in order to evaluate **h**, Maple needs to know if the argument is less than 1 or not, and this is impossible with the free variable **x**.

The following examples show correct ways of obtaining plots of the expression **f** and the functions **g** and **h**, on both the default interval $-10 \leq x \leq 10$ and the interval $-2 \leq x \leq 2$.

```
> plot(f,x);
> plot(f,x=-2..2);
> plot(g);
> plot(g,-2..2);
> plot(h);
> plot(h,-2..2);
```

Note that, if you open the **Style** menu and choose **Point** in the last plot, the discontinuity at $x = 1$ is more apparent.

The correct command to plot h on the interval $-.2 \leq x \leq .2$ is

```
> plot(h,-0.2..0.2);
```

or

```
> plot(h, -.2 .. .2);
```

but not

```
> plot(h,-.2...2);
```

since the last period is not interpreted as a decimal point.

It is correct to plot **g** with the statement

```
> plot(g(x), x = -2..2);
```

because **g(x)** is an expression, as indicated above. On the other hand, the following are both incorrect, since they combine the syntax for functions and expressions.

```
> plot(g(x), -2..2);

Warning in iris-plot: empty plot

> plot(g, x = -2..2);
```

Finally, the following shows that we can't plot **h** by converting it to the expression **h(x)**, since **h(x)** isn't an expression, as we pointed out above.

```
> plot(h(x), x = -2..2);

Error, (in h) cannot evaluate boolean
```

11.3 Re-executing Statements to Avoid Retyping Them

Re-executing statements can lead to errors because the Maple environment may not be what you expect when statements are re-executed. For example, suppose you calculate the areas of circles of radii 2 and 3 by entering

```
> r := 2;
```
$$r := 2$$

```
> A := Pi*r^2;
```
$$A := 4\pi$$

```
> evalf(%);
```
$$12.56637062$$

```
> r := 3;
```
$$r := 3$$

If you now re-execute the **evalf** command, you don't get the correct answer.

```
> evalf(%);
```

$$3.$$

The percent refers to the last output which was 3, not 9π. You might try entering **evalf(A)**, but this isn't correct either, since **A** hasn't been recalculated. The remedy is to first re-execute the statement where **A** is calculated, and then re-execute the **evalf** command.

Tip: Putting several statements on one line ensures that they will all be re-executed when the line is re-entered.

```
> r := 2;   A := Pi*r^2;  evalf(%);
```

$$r := 2$$

$$A := 4\pi$$

$$12.56637062$$

Now when **r := 2** is changed to **r := 3** and the line is re-executed, the **evalf(%)** command will work as you want it to. Misusing dittos is a very common mistake, and so caution is strongly advised.

Tip: Another way to avoid retyping is to use **Copy** and **Paste** from the **Edit** menu.

As another example of re-executing a statement, suppose you have entered the statement

```
> sum(1/(k*(k+2)), k = 1..10);
```

and later in the session you have assigned a value to **k**,

```
> k := 3;
```

Still later, you want to calculate the above sum but with upper limit 20, and so you change the 10 to 20 and re-execute.

```
> sum(1/(k*(k+2)), k = 1..20);

Error, (in sum) summation variable previously assigned,
            second argument evaluates to, 3 = 1 .. 20
```

This error is easily corrected by unassigning **k** and making it a free variable again.

```
> k := 'k';
```
$$k := k$$

The following error is similar, but not as easy to diagnose. Suppose **x** is a free variable (i.e., it has not been assigned a value), and you execute the statement

```
> int(x^2,x);
```
$$\frac{1}{3} x^3$$

Now suppose **x** is given a value

```
> x := 1;
```
$$x := 1$$

Then, at some later time, you decide to re-execute the **int** command.

```
> int(x^2,x);
```

```
Error, (in int) wrong number (or type) of arguments
```

Again, the remedy is to unassign **x**.

```
> x := 'x';
```
$$x := x$$

11.4 Forgetting to Save a Maple Session before It's Too Late

It sometimes happens that you want to interrupt a Maple session for one reason or another. For example, a computation may take longer than you anticipated. If you aren't able to save the session, you will lose work and be forced to start all over. For this reason, it's good practice to regularly open the **File** menu and save the session to a temporary file. For example, suppose you have been working for some time and have generated quite a number of lines in a Maple session. You want to calculate $\sum_{n=0}^{200} n!$, but you mistakenly enter

```
> sum(n!, n=0..2000);
```

This will probably take longer than you want to wait, so you try to save the session and exit, but Maple won't let you, and you feel annoyed for neglecting to first save the session.

11.5   Trouble Using On-Line Help

Frustration with the on-line help is a common complaint of new users, even to the point that they stop trying to use it. This is unfortunate, because the on-line help is an extremely valuable resource in Maple, whether used as reference for looking up some forgotten syntax, or as a vehicle for exploring the myriad commands and mathematics Maple puts at your disposal. We will look at some typical situations.

a. Suppose you want to plot two graphs on one coordinate system (you have seen it done before, so you know it's possible), but you don't remember the exact syntax. To see examples of the **plot** command, you enter

```
> ???plot
```

It shouldn't take you very long to find the following example.

```
> # multiple plots (in a set)
> plot({sin(x), x-x^3/6}, x=0..2);
```

If you need more explanation of the command, enter

```
> ?plot
```

for the full on-line help entry on **plot**, or else

```
> ??plot
```

for an abbreviated entry on just the syntax of applying the command.

b. Sometimes the explanations contain so much jargon and unfamiliar terminology that you're overwhelmed. It may be best to just ignore what you don't understand and not worry about it; you'll either learn it later, or else find you don't really need to know it. For example, suppose you want to find out about the **do** command, and you enter

```
> ?do
```

You immediately encounter terms like "statement sequence" and "boolean expression", which may not be too meaningful to you, so you skip to the examples (which you could have gone to immediately with the **???do** statement), and find that these are quite sufficient for your purposes. In fact, if you're so inclined, you can use the examples to help you understand the explanation and its unfamiliar terms.

c. Maple on-line help generally doesn't attempt to teach the mathematics that you may not understand, so you have to learn to separate the mathematics topics from the Maple topics. For example, the entry

```
> ?int
```

will tell you about Maple's **int** command for evaluating *definite integrals* and *indefinite integrals*, but it expects you to already know what these mathematical terms mean.

d. Sometimes something unexpected comes up, but it can safely be ignored, as in the following example. You want to solve the equation $e^x + x - 2 = 0$, and so you enter the following

```
> solve(exp(x) + x - 2=0,x);
```
$$-W(e^2) + 2$$

"What in the world is **W**?" you say to yourself. To find out, you decide to enter

```
> ?W
```

and see a screen full of information about something called Lambert's W function. Since you really only need a numerical approximation to the solution, you recall that you could have used **fsolve** instead of **solve**, and so you modify the above command and enter

```
> fsolve(exp(x) + x - 2=0,x);
```
$$.4428544010$$

The point is that since there is so much information available in the on-line help, you should learn to selectively ignore what you don't need, while hunting down what you do need.

Tip: The Keyword Search facility in the **Help** menu is useful for locating the particular help entry you need.

11.6   Forgetting to Include Library Packages

Some Maple commands reside in library packages that must be loaded into a Maple session before they can be used. To see a listing of the available library packages, enter

```
> ?packages
```

Sometimes you may remember a command and its syntax, but have forgotten that it is part of a package. If you try to use the command before loading the package, not only won't it work, but the reason may not be readily apparent to you. For example, suppose you want to plot some points, along with the graph of a function given by an expression, and you remember that the **display** command is one way to do this.

```
> p1 := plot([ [-1,1.1],[ .3,2],[ 1,2], [1.9,3]] , style=point):
> p2 := plot( x^2, x=-2..2 ):
> display({p1,p2});
```

The result isn't what you expected, just a lot of numbers. The remedy is to re-execute the above commands after you have entered

```
> with(plots):
```

When **?display** is entered, Maple indicates that **display** is in the **plots** library package.

```
> ?display
```

```
Try one of the following topics:
```
$$\left\{ simplex_{display}, plots_{display} \right\}$$

To get help on this topic, enter **?plots[display]**.

As another example, suppose you want to calculate the determinant of a certain 4×4 matrix, and so you enter

```
> A := matrix( [ [1,0,3,4], [2,3,4,0], [-1,4,5,6], [0,6,2,8] ] );
>
```
$$A := \mathrm{matrix}([[1,0,3,4],[2,3,4,0],[-1,4,5,6],[0,6,2,8]])$$

```
> det(A);
```
$$\det(\mathrm{matrix}([[1,0,3,4],[2,3,4,0],[-1,4,5,6],[0,6,2,8]]))$$

You won't get your answer unless you first enter

```
> with(linalg):
```

and then re-execute the previous two commands. This is because **matrix** and **det** are part of the linear algebra package **linalg**. Again, you would be alerted to this fact if you had entered **?matrix** or **?det**.

11.7 Confusion Between Exact and Approximate Calculations

Many commands work very differently depending on whether Maple is computing in exact arithmetic or not. A decimal point in a single number is enough to cause Maple to use approximate arithmetic. Consider the following examples.

a. Some commands will not work with approximate arithmetic. One such command is **factor**.

```
> factor(x^2 - 1);
```
$$(x-1)(x+1)$$

```
> factor(x^2 - 1.);
```

Error, (in factor/factor) floats not handled

Polynomials of degree four or less can be solved exactly

```
> solve(x^4 + x - 2=0,x);
```
$$1, -\%1^{1/3} + \frac{2}{9}\frac{1}{\%1^{1/3}} - \frac{1}{3},$$

$$\frac{1}{2}\%1^{1/3} - \frac{1}{9}\frac{1}{\%1^{1/3}} - \frac{1}{3} + \frac{1}{2}I\sqrt{3}\left(-\%1^{1/3} - \frac{2}{9}\frac{1}{\%1^{1/3}}\right),$$

$$\frac{1}{2}\%1^{1/3} - \frac{1}{9}\frac{1}{\%1^{1/3}} - \frac{1}{3} - \frac{1}{2}I\sqrt{3}\left(-\%1^{1/3} - \frac{2}{9}\frac{1}{\%1^{1/3}}\right)$$

$$\%1 := \frac{47}{54} + \frac{1}{18}\sqrt{249}$$

or approximately

```
> solve(x^4. + x -2=0,x);
```
$$1., -1.353209964, .1766049820 - 1.202820819\,I,$$
$$.1766049820 + 1.202820819\,I$$

On the other hand, polynomials of degree five or higher can't in every case be solved exactly, and here's how Maple responds when asked for exact solutions to such equations.

```
> solve(x^5-x = 0,x);    # this is easy to solve,
> #since it can be readily factored
```
$$0, 1, -1, I, -I$$

```
> solve(x^5 + x -3 = 0,x);    # this one is not so easy to solve
```
$$\mathrm{RootOf}(\_Z^5 + \_Z - 3)$$

However, Maple has a variety of ways for finding approximate roots, some of which we give below. Recall that a fifth degree polynomial has five roots, counting multiplicities, and that the complex roots of a real polynomial come in complex-conjugate pairs. The command **allvalues** can be used in conjunction with **RootOf**, which **solve** returns when it can't solve an equation exactly, as above.

```
> allvalues(%);
```
$$-1.041879540 - .8228703381\,I, -1.041879540 + .8228703381\,I,$$
$$.4753807567 - 1.129701725\,I, .4753807567 + 1.129701725\,I,$$
$$1.132997566$$

The following command returns only one approximate solution (notice the decimal point in the 3.)

```
> solve(x^5 + x - 3.=0,x);
```
$$-1.041879540 - .8228703381\,I$$

while **fsolve** will return all the real roots (since $x^5 + x - 3$ is a polynomial)

```
> fsolve(x^5 + x - 3=0,x);
```
$$1.132997566$$

If all the roots of a polynomial—complex as well as real—are desired, the following will always work.

> fsolve(x^5 + x -3=0,x,complex);

$$-1.041879540 - .8228703381\,I, -1.041879540 + .8228703381\,I,$$
$$.4753807567 - 1.129701725\,I, .4753807567 + 1.129701725\,I,$$
$$1.132997566$$

The command **evalf** can also be used, but only after **solve** has first been used (notice that the following result is the same as we obtained above using **solve** with a decimal number).

> evalf(solve(x^5 + x -3=0,x));
$$-1.041879540 - .8228703381\,I$$

Finally, we recall the example from Section 11.5, which also exhibits the difference between exact and approximate calculations.

> solve(exp(x)+x-2=0,x);
$$-W(e^2) + 2$$

> solve(exp(x)+x-2.=0,x);
$$.442854401$$

11.8 Trying to Get Maple to Do Too Much

It frequently happens that, when students are given an assignment to use Maple on a problem, they proceed under the assumption that they must use Maple for every step in the problem. This can lead to the absurd situation of having to painstakingly cajole Maple into performing steps that are completely obvious, and should just be done by hand. As an illustration of this point, we consider the following example. Suppose we are told that the temperature $T = T(t)$ of a cooling cup of coffee, where t denotes time, satisfies the equation $\ln(T - 70.) = -kt + C$, where k and C are unknown constants. We are also told that the temperature at $t = 0$ is 210 and that $T(1) = 190$. We want to solve for $T(t)$, and also for the time $t = t_1$ when $T(t_1) = 150$. The following statements show one way of having Maple solve this problem, but most beginning Maple users would probably choose to do some of the steps themselves. For example, it is a simple observation that $C = \ln(210. - 70.) = \ln(140)$. It is easier for the user to make this simplification than to enter the appropriate commands to force Maple to make the simplification.

```
> eq := ln(T-70.) = -k*t + C;
```
$$eq := \ln(T - 70.) = -k\,t + C$$

```
> params := solve( {subs({T=210,t=0},eq),
>   subs({T=190,t=1},eq)},{C,k});
```
$$params := \{\, k = .1541506800, C = 4.941642423 \,\}$$

Note: Although it appears that we have solved for k and C, the following shows that the numerical values haven't been assigned to k and C.

```
> k; C;
```
$$k$$

$$C$$

The numerical values must be assigned to k and C. Either enter

```
> k:=.15415068:
> C:=4.941642423:
```

or enter

```
> k := rhs(params[1]); C := rhs(params[2]);
```
$$k := .1541506800$$

$$C := 4.941642423$$

The following commands complete the problem.

```
> subs(T=150,eq);
```
$$\ln(80.) = -.15415068\,t + 4.941642423$$

```
> t1 := solve( %, t);
```
$$t1 := 3.63031671$$

11.9 Making Mistakes When Entering a Procedure

Maple procedures are similar to Fortran subprograms, Pascal procedures, or functions in C. They give a user a great deal of power and flexibility in customizing the Maple language.

The following example (which is one way of defining the function **h** used in Section **11.2**) illustrates the use of the **proc** and **end** keywords for writing a procedure.

```
> h := proc(x)
> if x < 1 then x^2 + 1 else 3*x fi;
> end;
h := proc(x) if x  <  1 then x^2+1 else 3*x fi end
```

Notice that there is no semicolon at the end of the first line. Also, observe that there is no output from Maple until the **end** statement is entered. This is because the words **proc**, **end**, and everything in between are considered *just one statement*. In other words, once **proc** is entered, Maple is not in the usual interactive mode until the **end** is encountered. We will look at two common errors when entering procedures. First, enter the first line above, and then enter the second line but with $x < 1$ erroneously typed as $x > 1$. Then, before the **end;** is entered, you notice the error, correct it, press ⟨ENTER⟩ again, and then enter the **end;** statement. Notice the output isn't what you intended.

```
h := proc(x) if 1 < x then x^2+1 else 3*x fi; if x < 1
then x^2+1 else 3*x fi end
```

The remedy is to reenter all three lines of code.

Another common error is to forget the **end** statement altogether and continue entering statements, which, of course, Maple treats as part of the procedure. A clue that something is wrong is that these new statements have no output. That might escape your notice, however, especially if the new statements are another procedure, which isn't supposed to have any output. For example, enter the following.

```
> h := proc(x)
> if x<1 then x^2 + 1 else 3*x fi;
> f := proc(y)
> y+1;
> end;
```

There is no output, since Maple is still processing the **h** procedure. The remedy in such a situation is to enter successive **end;**'s, until the result is

```
> end;

syntax error:
end;
     ^
```

and then reenter all the corrected statements, from the beginning.

Tip: If a procedure doesn't appear to be working correctly, you can obtain a listing with the **eval** command. For example, the statement

```
> eval(h);
```

will show the current version of **h**.

Remark: On most computers, <SHIFT> <ENTER> will give a new line without executing the Maple code. This is useful for entering multiline commands. Then, when <ENTER> is pressed without <SHIFT>, all of the lines will be executed.

11.10   Forgetting od after do and fi after if

These two errors are similar to leaving off the **end** after **proc**. In the following example, **fi** is omitted.

```
> h := proc(x)
> if x<1 then x^2+1 else 3*x;
> end;
```

Remember that, once you notice the missing **fi** in the second line and put it in, you must re-enter the entire procedure.

Tip: As soon as you type **if**, skip a space and type **fi**; then back up and type what goes in between. This may help you to remember the **fi**.

The following example shows the correct use of the **do** and **od** keywords:

```
> s := 0:
> for i to 10 do
> s := s+i:
> od:
> print(s);
```

55

Observe that the last two statements produce an output. Now enter these statements again, only without the **od:** statement. Notice that Maple has stopped producing any output. Anytime there is no output, you should be alerted to the possibility that you have forgotten to end a compound statement, such as **do..od**, **proc..end**, or **if..fi**. The earlier tip is also valid for **do..od**: type **od** right after **do**; then back up and insert what goes in between them.

Chapter 12

Projects

12.1 Parameterizing Letters

Calculus Prerequisites: None.

Background: A curve of the form $x = f(t)$ and $y = g(t)$ where t is a variable ranging over an interval $a \leq t \leq b$ is called a *parameterized curve* (with t as the parameter). The goal of this problem is to use parameterized curves to form letters.

Letters can be made with a combination of straight line segments, circles and ellipses. These curves can be described using parameterized curves as in the following examples.

1. The straight line segment from the point $P(x_1, y_1)$ to the point $Q(x_2, y_2)$ is parameterized by $x = x_1 + t(x_2 - x_1)$ and $y = y_1 + t(y_2 - y_1)$. For example, if $P = (1, 0)$ and $Q = (2, 3)$, then the following Maple commands plot the line segment between these two points.

   ```
   > x1 := 1: y1 := 0: x2 := 2: y2 := 3:
   > x := x1 + t*(x2-x1):  y := y1 + t*(y2-y1):
   > plot([x,y,t=0..1], 0..3,0..3,scaling=constrained);
   ```

   Recall that the **0..3, 0..3** syntax refers to the viewing window $0 \leq x \leq 3$ and $0 \leq y \leq 3$. This can be changed to any window that includes the desired plot. The **scaling = constrained** syntax forces the scales on the x and y axes to be the same.

2. An arc of a circle centered at (a, b) of radius r can be parameterized by $x = a + r\cos(t)$ and $y = b + r\sin(t)$. For example, the following Maple commands parameterize a three-quarter circle of radius 2 centered at $(-1, 3)$.

```
>   x :=-1+2*cos(t): y := 3+2*sin(t):
> plot([x,y,t=0.. 3*Pi/2], -4..2,-1..5,scaling=constrained);
```

3. Similarly, an arc of an ellipse, centered at the point (a, b) with horizontal axes radius r and vertical axes radius s is parameterized by $x = a+r\cos(t)$ and $y = b + s\sin(t)$.

4. The following commands plots a small rectangle.

```
> u[1] := 1+.5*t: v[1] := 1:  line1:=plot([u[1],v[1],t=0..1],0..2,0..2):
> u[2] := 1: v[2] := 1 - .5*t: line2:=plot([u[2],v[2],t=0..1],0..2,0..2):
> u[3] := 1 + .5*t: v[3] := .5: line3:=plot([u[3],v[3],t=0..1],0..2,0..2):
> u[4] := 1.5: v[4] := 1-.5*t: line4:=plot([u[4],v[4],t=0..1],0..2,0..2):
> with(plots):
> display([line1,line2,line3,line4],scaling =constrained);
```

Alternatively, **u[i]**, **v[i]** for $i = 1..4$ can be defined as above and then the plot process can be automated with a **for** loop, as in the following sequence of commands.

```
> rect := NULL: #this clears the variable rect
> for i from 1 to 4 do #start the loop
>      rect := rect , [u[i],v[i],t=0..1]: #string segments together
> od:
> plot({rect},0..2,0..2,scaling=constrained):
```

Assignment:

1. Plot the three parts of a capital letter B, using a straight line segment and either two semicircles, or two semi-ellipses.

2. Repeat Exercise 1., but with the letter moved .5 units above the x axis and .5 units to the right of the y axis, and with its size doubled.

3. Use parametric curves to make graphs of the letters in your name.

Follow-up Activity: The Laboratory Project: *Bezier Curves* at the end of Stewart's Section 3.5.

12.2   A Power Relay Station

Calculus Prerequisites: Setting up word problems and graphing functions. Read Chapter 1 in Stewart's **Calculus**.

Assignment: On a long straight highway, there are 15 communities located at mile markers 1, 1.5, 2, 2.35, 5, 5.1, 6.35, 6.45, 6.5, 7.33, 9.6, 10.11, 12.5, 14.7, 16. The electric company which serves these 15 communities would like to construct a power relay station somewhere along this highway. The company would then run separate power lines from the relay station to each of the 15 communities. To minimize the cost of constructing these 15 power lines, the company would like to locate the relay station so as to minimize the sum of the distances to each station.

Answer the following questions.

1. Where along the highway should the electric company build its relay station? Justify your answer. Is this location unique?

2. Suppose a 16th community, located at mile 20, is added to the highway. Now where should the electric company locate its relay station? Is this location unique?

Hint: Think of the communities as located along the x-axis at the mile markers that are given. Suppose the relay station is located at position x. Write the sum of the distances from the power station to the communities in terms of x.

12.3 Speed Limits

Calculus Prerequisites: average velocity, limits. Read Sections 2.1, 2.2, 2.3 and 2.6 in Stewart's **Calculus**.

Background: The goal of this project is to learn about average and instantaneous velocities. In the process you will learn to use Maple to define, evaluate, plot, and take limits of functions. You will use the Maple commands that define, evaluate, and plot functions. In addition, you will use the Maple commands to compute limits at finite points and at infinity. Before you start the project, be sure you try each of the following examples.

- Definition and evaluation of functions. The following Maple command defines the function $x(t) = \frac{t^3 - 9t}{3t^3 - 9t^2 + t - 3}$, which might be interpreted as the position of an object at time t.

```
> x := t -> (t^3 - 9*t)/(3*t^3 - 9*t^2 + t - 3);
```

Then the position at $t = 1$ is

```
> x(1);
```

You can also define functions of two or more variables. For example, the following gives the average velocity of the object between the times $t = t_1$ and $t = t_2$.

```
> AveVel := (t1,t2) -> (x(t2) - x(t1))/(t2 - t1);
```

Then the average velocity between $t = 1$ and $t = 2$ is

```
> AveVel(1,2);
```

- Plots. The function $x(t)$ defined above, can be plotted on the interval $0 \leq t \leq 10$ by using the Maple command

```
> plot(x(t),t=0..10);
```

- Limits. Notice that the function $x(t)$ is undefined at $t = 3$ because the denominator is zero there. Maple agrees.

```
> x(3);
Error, (in f) division by zero
```

However, Maple can compute the limit as t goes to 3.

```
> Limit(x(t),t=3);   value(%);   evalf(%);
```

Maple can also compute the limit as t goes to ∞.

```
> Limit(x(t),t=infinity);   value(%);
```

You should be able to verify these limits, both in the plot and by algebraically manipulating the formula for $x(t)$.

Assignment: Answer the following questions on the lab report form.

1. A Texas Aggie suddenly decided to take his girlfriend to Houston for the day. So he hopped in his car and started to drive from College Station to Houston. When he got to Navasota, he suddenly realized that he had forgotten his girlfriend. Embarrassed, he slowly drove back to College Station and got his girlfriend. All this took 2 hours. Then they proceeded to Houston, taking another 2 hours. His distance (in miles) from College Station at time t (in hours) was $x(t) = \frac{195t^2(t-2)^2}{2+8t^2}$.

   (a) Enter his position function $x(t)$ into Maple using an arrow definition.
   Maple Procedure: *Don't forget your parentheses and check your output.*

(b) Plot his position $x(t)$ for times between $t = 0$ when he first left College Station and $t = 4$ when he finally arrived in Houston.

(c) Approximately when did he arrive in Navasota? Approximately, what is the distance from College Station to Navasota?

Maple Procedure: *Click in the plot window at the first maximum and read off the t and x coordinates. Note: Maple calls the horizontal axis X and the vertical axis Y when they should really be t and x*

(d) What is the exact distance from College Station to Houston?

Maple Procedure: *Compute x(4).*

Maple Procedure: *In the following problems use the Maple function of two variables, AveVel, given above to compute the average velocities.*

(e) Find his average velocity

    i. from College Station to Navasota, the first time out. (approximate).

    ii. from Navasota back to College Station. (approximate).

    iii. from College Station to Houston, the second time out. (exact).

    iv. from College Station to Houston, the whole trip. (exact).

Maple Procedure: *To avoid round-off error in the next problem, execute*

```
> Digits := 20;
```

(f) Find his average velocity between time $t = 3$ and time $t = 3 + h$ for the following values of h

1	.1	.01	.0001
-1	$-.1$	$-.01$	$-.0001$

(g) Use the values calculated above to guess his instantaneous velocity at $t = 3$.

Maple Procedure: *Before the next problem, be sure you clear h by executing*

```
> h := 'h';
```

(h) Compute his average velocity between time $t = 3$ and time $t = 3 + h$ for a variable value of h. Use this formula to compute his instantaneous velocity at $t = 3$ by computing the limit

$$\lim_{h \to 0} AveVel(3, 3 + h)$$

(i) Recompute the limit in (h) by algebraically manipulating $AveVel(3, 3+ h)$ using the **expand** and/or **factor** commands until the h in the denominator cancels. Then substitute in $h = 0$ using the **subs** command.

(j) *10% extra credit.* At what time does he pass through Navasota on his second time out?

   Maple Procedure: *If X is the distance to Navasota, solve the equation $x(t) = X$ using Maple's **fsolve** command.*

2. A parachute jumper's velocity is $v(t) = \frac{98t + 100t^2}{1 + 20t^2}$ in meters per second where t is in seconds.

   (a) Enter her velocity function $v(t)$ into Maple using an arrow definition.

   Maple Procedure: *Don't forget your parentheses and check your output.*

   (b) Plot her velocity $v(t)$ for the first 5 seconds of her fall and again for times between $t = 0$ (when she jumps) and $t = 120$ seconds (when she lands).

   (c) Find an approximate value for her maximum velocity.

   Maple Procedure: *Click in the plot window on the maximum. Remember that in the plot window, X means t and Y means v.*

   (d) Find an exact value for her terminal velocity by computing $\lim_{t \to \infty} v(t)$.

   Note: You should be able to verify this limit, both in the plot and by algebraically manipulating the formula for $v(t)$.

   (e) Her landing will be safe if her velocity is less than 6 meters per second. Was her landing safe?

12.4   The Ant and the Blade of Grass

Calculus Prerequisites: tangent lines. Read Sections 2.6 and 3.1 in Stewart's **Calculus**.

Background: In this project, you will learn how to use Maple to differentiate a function and construct a tangent line, to plot a series of points connected by line segments, to plot several things on the same graph, and to solve an equation. You will need to differentiate functions using **D**, construct a tangent line, plot points along with functions, and solve equations using **fsolve**. Before you start this project, be sure you try each of the following examples.

- Differentiating functions.
  If a function f is defined using an arrow definition, for example

```
> f := x -> x^3;
```

then its derivative is computed using Maple's **D** operator as follows.

```
> D(f);
```

Notice that the output is an arrow-defined function but it doesn't have a name. If you wish to give it a name, say Df, then you type

```
> Df := D(f);
```

- Tangent lines.
  If you want to find the line tangent to the curve $y = f(x)$ at the point $(a, f(a))$, you first find the slope $m = f'(a)$ and then use the point-slope equation of the line: $y - f(a) = m(x - a)$. For the function $f(x) = x^3$ discussed above, the slope of the tangent line at $x = 2$ is found from

```
> a := 2;   m := Df(a);
```

and the equation of the tangent line is

```
> y - f(a) = m*(x-a);
```

Notice that this is an equation and not an assignment (there is no :=). So nothing has been stored in memory. By solving the equation for y, i.e., $y = m(x - a) + f(a)$, you can also define the tangent line as a function.

```
> ftan := x -> m*(x-a) + f(a);
```

To see its formula, type

```
> ftan(x);
```

Hint: The benefit of defining **a:= 2;** instead of just typing 2 each time, is that if you need to compute the tangent line at several points, you can cut and paste the whole sequence of commands and only change the value of a before re-executing..

- Plotting several functions.
  The function $f(x)$ and its tangent line $f_{\tan}(x)$, defined above, can be plotted together on the interval $-5 \le x \le 5$ by enclosing the two functions in curly braces **{ }** and using the Maple **plot** command.

```
> plot({f(x),ftan(x)},x=-5..5);
```

- Lists and plotting points.
  In Maple an ordered list is denoted by separating the items by commas

and enclosing the list in square brackets **[ ]**. Thus, the point $(2,5)$ is entered as the ordered pair **[2,5]**, while a list of points might be

```
> octagon := [ [1,0], [4,0], [5,1], [5,4], [4,5], [1,5], [0,4], [0,1], [1,0] ];
```

If you plot this list of points and connect the dots, you should get an octagon.

```
> plot(octagon);
```

If you have several lists of points, enclose them in curly braces **{ }**. Try the following.

```
> line1 := [ [1,3], [2,3] ];   line2 := [ [3,3], [4,3] ];
> you := [ [1,2], [2,1], [3,1], [4,2] ];
> plot({octagon, line1, line2, you});
```

What did you get?

Finally you can even plot lists of points together with functions and also specify both the x-range and the y-range. For example

```
> plot({f(x), ftan(x), octagon, you}, x=0..5, y=0..20);
```

- Solving equations.
  Suppose you want to solve the equation $x/\pi + \sin(x) = 1$. You first enter the equation to make sure you have typed it properly.

```
> eq := x/Pi + sin(x) = 1;
```

You can then plot the left hand and right hand sides of this equation, to see where they intersect. Notice the use of the commands **lhs** and **rhs**.

```
> plot({ lhs(eq), rhs(eq) }, x=-2*Pi..4*Pi);
```

So there are three intersection points, i.e., solutions, one between 0 and 2, one between 2 and 4, and one between 4 and 6. To find the solutions, you first try Maple's **solve** command.

```
> solve(eq,x);
```

This finds the exact solution between 2 and 4. The other two solutions can't be found exactly. So you use Maple's **fsolve** command to find approximate solutions.

```
> fsolve(eq,x);
```

Notice that this is the same solution. So you need to add a range to the
fsolve command.

```
> fsolve(eq,x=0..2);   fsolve(eq,x=4..6);
```

We now have all three solutions.

Assignment: An ant is walking (to the right) over its ant mound, whose height
(in inches) is given by the function: $h(x) = \frac{x^2/16 - 2x + 80}{(x^2/16 - 2x + 20)^2}$. Nearby there is a
blade of grass, which is located as the line segment from $(32, 1/5)$ to $(32, 8)$.
The goal in this lab is to find the point where the ant first sees the blade of
grass. You can assume that the ant's line of sight is the tangent line to the ant
mound. The following series of questions will lead you to the solution. Answer
them on the lab report form.

1. Define the function **h** that gives the height of the ant mound, using an
   arrow definition. Define the line segment occupied by the blade of grass
   and name it **grass**. Plot the ant mound and the blade of grass on the same
   graph.

2. Compute the derivative of **h** and name it **Dh**. On the lab report, record
   your Maple input but not the output.

3. Compute the tangent line to $y = h(x)$ at $x = 12.5$ and define it as a
   function **htan** using an arrow definition. Plot the ant mound, the blade of
   grass, and the ant's line of sight when the ant is at $x = 12.5$. Can it see
   the blade of grass? Find the height H where the tangent line crosses the
   line $x = 32$ by evaluating **htan(32)**.

4. Compute the tangent line to $y = h(x)$ at $x = 15.5$ and define it as a
   function **htan** using an arrow definition. Plot the ant mound, the blade of
   grass, and the ant's line of sight when the ant is at $x = 15.5$. Can it see
   the blade of grass? Find the height H where the tangent line crosses the
   line $x = 32$ by evaluating **htan(32)**.

5. We can now see that, when the ant is at some position $x = a$ between 12.5
   and 15.5, it can first see the top of the blade of grass. We want to find a.
   So, compute the tangent line to $y = h(x)$ at $x = a$ for a variable a. Define
   the tangent line at $x = a$ as a function **htan** using an arrow definition.

   Note. If you previously gave **a** a value, clear it by executing

```
> a:='a';
```

6. You can no longer plot the tangent line because its formula contains a variable, namely a. However, you can still find the height H where the tangent line crosses the line $x = 32$ by evaluating **htan(32)**. When this height H equals the height of the blade of grass, the ant can just begin to see the blade of grass. Use Maple's **fsolve** command to solve for the value of a where H equals the height of the blade of grass. (You may need to specify a range for a in the **fsolve** command.) Denote the solution by **A**.

7. For the value A found in problem 6, compute the tangent line to $y = h(x)$ at $x = A$ and define it as a function **htan** using an arrow definition. Plot the ant mound, the blade of grass, and the ant's line of sight when the ant is at $x = A$. Can it see the blade of grass? Find the height H where the tangent line crosses the line $x = 32$ by evaluating **htan(32)**.

8. There is a second solution to the equation $H = 8$. What is wrong with this solution?

12.5   Tangent and Normal Lines

Calculus Prerequisites: Computing tangent and normal lines to curves. The second question posed in this project is considerably more difficult than the first. Read Sections 2.6 and 3.1 in Stewart's **Calculus**.

Assignment: Answer the following two questions.

1. Show that every point (x, y) with $y < x^2$ has the property that it belongs to two distinct tangent lines to the curve $y = x^2$.

2. Find the equation(s) of the curve C with the property that each point on C belongs to two distinct normal lines to $y = x^2$.

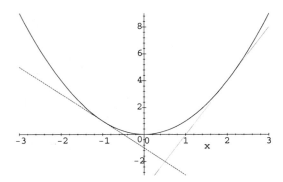

Hints: For the first question, if (a, b) belongs to the tangent line to $y = x^2$ at (x, x^2), then x is the solution of a quadratic equation obtained by setting the slope of the chord from (x, x^2) to (a, b) equal to the derivative of $y = x^2$ (draw a picture). The analogous equation for the second question, involving normal lines, is of the form $f(x) = 0$ where $f(x)$ is a cubic polynomial. In order for a cubic equation to have exactly two roots, one of them, say $x = c$, must also be a root of the derivative of f. Therefore in order for (a, b) to belong to a normal line to $y = x^2$, the resulting equations, $f(x) = 0$ and $f'(x) = 0$, must both have a root at the same point (namely, $x = c$).

12.6 Center of Gravity

Calculus Prerequisites: computing normal lines to a curve and simple integration. Read Section 3.1 and Chapter 5 in Stewart's **Calculus**.

Background: The center of mass of a region bounded above by $y = f(x)$ and below by $y = g(x)$, $a \le x \le b$, is the point $(x0, y0)$ where

$$x0 = \frac{1}{A} \int_a^b x[f(x) - g(x)]\, dx$$

$$y0 = \frac{1}{A} \int_a^b 1/2 \left([f(x)]^2 - [g(x)]^2\right)\, dx$$

Consider a thin metal plate which is cut in the shape of a parabola bounded by the parabola $y = \frac{1}{40}x^2$ and the line $y = 10$. Suppose the x-axis corresponds to ground level and the y-axis corresponds to up and down. If the density of the plate is uniform, then the center of gravity will be located on the y-axis (at $(0,6)$). If the plate is rolled along the x-axis slightly, the plate will become "off-balance" and will try to roll to a new equilibrium state.

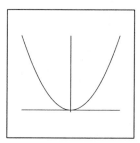 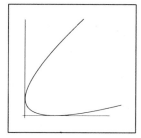

Assignment: Consider the thin metal plate described above. Suppose the plate is pushed very slightly so that it rolls in the positive x direction.

1. Give a geometric argument which explains why the plate will not return to its original position.

2. Find the new equilibrium state that the plate will move towards.

3. Give the coordinates of the center of gravity of the plate in its new equilibrium position.

4. Repeat problems 2 and 3 for the generic parabola $y = ax^2$ where $a > 0$ (still use the line $y = 10$).

5. Give a parametric plot (in terms of the parameter a) of the centers of gravity found in problem 4.

12.7 Calculus I Review

Calculus Prerequisites: limits, differentiation and integration, area under a curve. Read Chapters 1 - 5 in Stewart's **Calculus**.

Background: In this project, you will review the chain rule, differentiation, and integration, as well as the Maple commands for these operations.

Review the following notions before beginning this project: domain of a function, composition of functions, and the geometry of derivatives and integrals.

The symbols $f(x)$, $g(x)$, $h(x)$ and $k(x)$ will refer to the following functions

$$f(x) = \sqrt{x - 2}, \quad g(x) = x^2, \quad h(x) = f \circ g(x), \quad k(x) = g \circ f(x)$$

You will be doing various computations with these four functions. Enter these functions into a Maple session. The Maple commands to do this for the functions $f(x)$ and $g(x)$ are

```
> f:=x->sqrt(x-2); g:=x->x^2;
```

Note: One way of entering the composition of two functions is to use the symbol @. Thus, one could enter the function $k(x)$ with either of the following Maple commands.

```
> k:=g@f;
> k:=x->g(f(x));
```

To see the explicit algebraic expression of $k(x)$, for example, enter the Maple command

```
> k(x);
```

Assignment: Answer the following questions.

1. Using the Maple command **plot({f(x),g(x)},x=-7..7)**, plot the graphs of $f(x)$ and $g(x)$ for x between -7 and 7. Be sure to label the graphs.

2. The function g has all real numbers for its domain. What is the domain for the function f?

3. Explain why the plot for $f(x)$ is not drawn for all numbers between -7 and 7.

4. Plot the graphs of $h(x)$ and $k(x)$ in Maple as in problem 1. Be sure to label the graphs.

5. Explain why the straight line passing through the point $(0, -2)$ is not the graph of $k(x)$, and draw a better graph.

6. Without using Maple, compute the derivatives of the two functions $h(x)$ and $k(x)$. After you have done this, compare your answer with Maple's.

   Note: You can have Maple calculate derivatives with either the **D** operator or the **diff** operator.

   ```
   > deriv_k:=D(k)(x);
   > deriv_k:=diff(k(x),x);
   ```

7. The answer you get for the derivative of $k(x)$ should *not* be the same as Maple's. Why not?

8. Sketch the region bounded by the curves $y = 0$, $y = f(x)$, $y = g(x)$, and $x = 6$.

9. Calculate the area bounded by the curves you sketched in problem 8, by using the Maple commands

   ```
   > Int(g(x),x=0..2) + Int(g(x)-f(x),x=2..6);
   > value(%);
   ```

   Explain what each of the integrals represents.

   > *Note:* Instead of using the Maple commands **Int** and **value,** you can use the command **int**. If you do this Maple will not print out the integrals; only the answer will be printed. *Get in the habit of using commands that print out what you want calculated.* This makes it a lot easier to find and correct typos and other mistakes.

10. What happens when you enter the following Maple command?

    ```
    > int(g(x),x=0..2) + Int(g(x)-f(x),x=2..6);
    ```

11. Explain the differences in output between the following approaches to calculating the derivative of the function $\dfrac{1}{x(x-2)}$.

    (a) > **diff(1/x*(x-2),x);**
    (b) > **p:=(x*(x-2))^(-1); diff(p,x);**
    (c) > **Diff(1/x*(x-2),x); value(%);**

    Which values are correct? How can the others be fixed? After fixing them, which approach do you feel is best? Why?

12.8   Calibrating a Dipstick

Calculus Prerequisites: volumes of revolution. Read Sections 6.2 and 6.3 in Stewart's **Calculus**.

Assignment: Suppose the semi-circle $y = -\sqrt{100 - x^2}$, $-10 \leq x \leq 10$ is rotated about the y-axis to create a circular bowl (10 units deep). Construct a dipstick which when inserted vertically into the bowl will determine when the bowl is one quarter full; one-half full; and three-quarters full.

Now repeat the same problem assuming that the bowl is generated by rotating the curve $y = -(81 - x^4)^{0.4}$, $0 \leq x \leq 3$ around the y-axis. This time, the integrals must be evaluated numerically. A little trial and error will be needed to determine the level of the dipstick where the volume of the bowl is one-quarter full, one-half full, etc.

12.9   The Center of the State of Texas

Calculus Prerequisites: center of mass and numerical integration. Read Section 8.7 in Stewart's **Calculus**.

Background: The goal of this project is to compute (an approximation to) the center of mass of the state of Texas from data (given here) that represent the state's boundary.

The center of mass of a region bounded above by $y = f(x)$ and below by $y = g(x)$, $a \leq x \leq b$, is the point $(x0, y0)$ where

$$x0 = \frac{1}{A} \int_a^b x[f(x) - g(x)]\, dx$$

$$y0 = \frac{1}{A} \int_a^b 1/2 \left([f(x)]^2 - [g(x)]^2\right)\, dx$$

and A represents the area of the region. For the state of Texas, the origin will be located at the western tip (near El Paso) and the x-axis is the extension of the east-west border between New Mexico and Texas. In the above integral formulas, the graph of $y = f(x)$ represents the upper boundary and the graph of $y = g(x)$ represents the lower boundary of the region in question. Since there are no analytical formulas for f and g for the boundary of the state of Texas, approximations to the above integrals must be computed using the following data, which represent the boundary of the state. The second coordinate represents the values of f and g at integer values of x from $x = 0$ to $x = 11$, where each unit represents 69 miles.

```
> north:=[[0,0],[1,0],[2,0],[3,0],[3,4.5],[4,4.5],[5,4.5],[6,4.5],
> [6,2.2],[7,2.1], [8,1.8],[9,1.9],[10,1.8],[11,1.7],[11,-2.2]];
> south:=[[0,0],[1,-1.1],[2,-2.5],[3,-2.9],[4,-2.3],[5,-2.8],[6,-4.4],
> [7,-5.8],[8,-6.1],[9,-3.3],[10,-2.8],[11,-2.2]];
```

Note that there are two y-values given in the northern boundary for both $x = 3$ and $x = 6$ (because $x = 3$ and $x = 6$ represent the two north-south boundaries of the Panhandle).

Assignment: Use these data to compute the area A of the state of Texas using the trapezoidal rule. Then compute the center of mass of the state using the trapezoidal rule to compute the integrals. Include a plot that shows the boundary of the state with its center of mass.

Hints: Enter the above data as lists labeled *north* and *south* (or make up your own names). To refer to the entries in a list, use brackets []; for example, **south[3]** refers to the point $[2, -2.5]$ and **south[3][2]** refers to the second entry of this point (i.e., -2.5). For example, the following command sums the second entries (the y-values) of all the points on this list.

```
> Sum(south[i][2],i=1..12); value(%);
```

To plot the boundary of the state, enter the command

```
> plot({north,south});
```

To plot the state boundary along with the center point $[x0, y0]$, issue the following commands

```
> texas:=plot({north,south}):
> center:=plot([x0,y0],style=point):
> with(plots): display([texas,center]);
```

The first two commands assign the labels *texas* and *center* to the state boundary and the point [x0, y0], respectively (colons are used instead of semicolons in order to suppress the lengthy output of numerical garbage). The third line of commands displays both plots on the same plot window.

12.10   The Skimpy Donut

Calculus Prerequisites. Volume, surface area, and max/min theory. Read Sections 4.7 and 6.2 in Stewart's **Calculus**.

Background: The GETFAT Donut company makes donuts with a thin layer of chocolate icing. The company decides to cut costs by minimizing the amount of chocolate icing used in each donut without shrinking the size or weight of the donut. The problem, then, is to determine the dimensions of the donut of a fixed volume that minimize the surface area.

A donut has the shape of a torus which is determined by revolving a circle around a line, as shown in the diagram below.

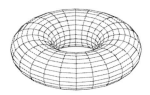

If the circle has radius b and the line is a distance b from the center of the circle, then the donut can be generated by revolving the circle $(x-a)^2 + y^2 = b^2$ around the y-axis. The volume of this donut can be found using the technique of cylindrical shells. The surface area can be found by writing x as a function of y and using the usual formula for the surface area of a solid of revolution (with x playing the role of the dependent variable and y - the independent variable).

Assignment. Answer the following questions.

1. Compute the volume V of the donut as a function of a and b. As a check, the volume of the donut should be $\dfrac{\pi^2}{2} \approx 5$ cubic inches when $a = 1$ inch and $b = 1/2$ inch.

2. Compute the surface area of the donut as a function of a and b. As a check, the surface area of the donut should be $2\pi^2$ cubic inches when $a = 1$ and $b = 1/2$ inch.

3. With the volume fixed at $\dfrac{\pi^2}{2}$ cubic inches, find the dimensions a and b of the donut which minimize the surface area of the donut. Note: for this problem, you will have to determine the range of allowable a and b.

4. Is there a maximum surface area for a given volume?

12.11   Search for the Meteor

Calculus Prerequisites: the definition of the hyperbola.

Assignment: The goal of this project is to solve the following problem. A meteor crashes somewhere in the hills that lie north of point A. The impact is heard at point A and, 6 seconds later, it is heard at point B. Three seconds still later it is heard at point C. Locate the point of impact of the meteor, given that A lies four miles due east of B and two miles due west of C. The speed of sound is roughly .20 miles per second.

   Solve the above problem. Use complete sentences to describe your procedure. Include an accurate plot with your explanation.

Hints: The equation of the hyperbola centered at the origin with focal points at $(+c, 0)$ and $(-c, 0)$ along the x-axis, is given by

$$\frac{x^2}{a^2} - \frac{y^2}{c^2 - a^2} = 1$$

Here, $2a$ represents the difference $|d_1 - d_2|$, where d_1 is the distance between an arbitrary point (x, y) on the hyperbola and the focal point $(+c, 0)$, and d_2 is the distance between (x, y) and the focal point $(-c, 0)$.

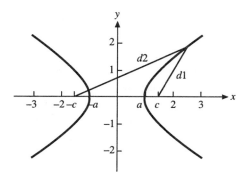

For this project, locate points A, B, and C on the x-axis. The choice of origin is somewhat arbitrary, but a convenient choice of origin is the midpoint between points A and B. From the information given, the meteor must lie on a hyperbola centered at the origin with points A and B as focal points. Find the equation of this hyperbola. Likewise, the meteor must also lie on a hyperbola with focal points at B and C. (Where is this hyperbola centered?) Thus, the meteor must lie on the point of intersection between the two hyperbolas.

Note that a hyperbola has two branches and you must take care to choose the correct branch. The information given in the problem should lead you to the correct branch.

12.12   Curves Generated by Rolling Circles

Calculus Prerequisites: parameterized curves, the cycloid and arc length. Read Section 11.3 in Stewart's **Calculus**.

Background: The goal of this project is to use Maple's plot and integrate commands to help answer questions on the cycloid, which is a curve generated by a point on a rolling circle. A more complicated version of the cycloid is also considered.

Consider a wheel of radius R. Fix a point on the rim of the wheel. Now let the wheel roll on level ground and consider the path traced out by the point P (see the figure). This path is called the cycloid.

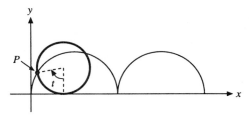

Assignment: Answer the following questions, using Maple where appropriate.

1. Show that the cycloid is parameterized by the formulas

$$x(t) = R(t - \sin(t)) \text{ and } y(t) = R(1 - \cos(t))$$

   Here, t is the angle between the vertical and the ray that extends from the center of the circle to P (so $t = 0$ when P is at the origin).

2. Use Maple to plot two arches of this cycloid with $R = 1$. To plot a parametric curve (x, y) (where x and y are expressions in t) over the interval $a \le t \le b$, use the command

   ```
   > plot([x,y,t=a..b]);
   ```

   Note the square brackets that are used with this plot command. Now unassign R (**R := 'R';**).

3. Compute the arc length of one arch of this cycloid (for a general value of R). Recall that the arc length of a parameterized curve $(x(t), y(t))$ for $a \le t \le b$ is given by

$$\int_a^b \sqrt{(x')^2 + (y')^2} \, dt$$

4. In the previous problem, you determined the arc length of one arch by computing the integral over the interval $0 \le t \le 2\pi$. Presumably, computing the appropriate integral over the interval $0 \le t \le 4\pi$ should give the arc length of two arches. However Maple computes this integral as zero, which makes no sense (try this). See if you can find where Maple is making an error. Hint: $\sqrt{u^2} \ne u$ when $u < 0$.

5. The slope of a parameterized curve $(x(t), y(t))$ is given by

$$\frac{dy}{dx} = \frac{dy/dt}{dx/dt}$$

Use Maple to show that the limit of the slope of the tangent line as $t \to 0^+$ is infinity.

Note: It may be easier to show that $1/\text{slope} \to 0$ as $t \to 0^+$.

6. Now suppose a circle of radius a rolls around the outside of the circle of radius $R > a$ centered at the origin (see the accompanying diagram). Find the parameterization $P = (x(t), y(t))$ that describes the path of a fixed point P on the rolling circle. Here t is the angle measured counterclockwise from the positive x-axis to the line segment that runs from the origin to the center of the rolling circle. Assume that P is located at the point $(R, 0)$ when $t = 0$. Compute the arc length of one of the arches of this path.

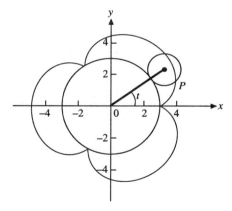

Related Activity: See the Laboratory Project: *Families of Hypocycloids* at the end of Section 11.1 in Stewart's text.

12.13 Gravitational Force

Calculus Prerequisites: surface area, and the ability to view an integral as a process of summation. Read Sections 5.1 and 9.2 in Stewart's **Calculus**.

Background: The goal of this project is to determine the gravitational force between a mass m that is concentrated at a point and other objects whose mass may be spread out over a large region—for example, a ring, a spherical shell, or a solid ball (such as the Earth).

This project assumes knowledge of integration and the area of a surface of revolution. The gravitational attraction between two bodies of mass m and M (which are concentrated at two points in space) is given by the following inverse square formula

$$G\frac{mM}{r^2}$$

where r is the distance between the two masses and G is the gravitational constant (whose value depends on the units involved). If one of the objects, say M, is spread out over a large region (such as the Earth), then the distance between m and various points of M will vary (i.e., there is no well-defined value for r). In order to determine the gravitational attraction between m and a large object M, the large object must be subdivided into smaller pieces whose gravitational attraction with m is easy to compute. Then the gravitational attractions of the smaller pieces must be summed (integrated) to determine the total gravitational attraction. Maple can help with the computations.

Assignment: Complete the following problems which will lead to the computation of the gravitational force between a point mass m and a large homogeneous spherical ball M. Use complete sentences to describe your solution.

1. *Surface area property of the sphere* (this result will be used later). Consider the sphere of radius R centered at the origin. Slice this sphere along two parallel planes. Show that the surface area of the slice only depends on the distance between the two planes (and not on their locations). *Hint:* Think of the sphere as the surface obtained by revolving a circle of radius R, centered at the origin, about the x-axis. Let the planes be perpendicular to the x-axis at $x = a$ and $x = b$, where $-R \leq a \leq b \leq R$. Now show that the surface area of revolution of this circle between $x = a$ and $y = b$ depends only on $b - a$. This problem can be done by hand or by using Maple.

2. Show that the magnitude of the gravitational attraction between the point mass m and a uniform circular wire of radius r and total mass M, whose center is a distance h from m, is

$$\frac{GmMh}{(h^2 + r^2)^{\frac{3}{2}}}$$

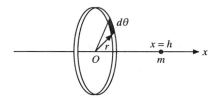

Hint: Divide the circular wire of mass M into angular sectors of angular width $d\theta$. Each sector has mass $\dfrac{M\,d\theta}{2\pi}$. By symmetry, only the component of the gravitational force that is parallel to the x-axis needs to be computed (since the component perpendicular to the x-axis will cancel with the perpendicular component of the analogous angular sector on the opposite side of the circle). Then add up (integrate) over $0 \le \theta \le 2\pi$. This integral is simple enough to do without Maple.

3. Show that the gravitational force between the point mass m and a hollow spherical shell of radius R and mass M is

$$\frac{GmM}{a^2}$$

where $a > R$ is the distance between the mass m and the center of the shell.

This means that the gravitational attraction is the same as if all the mass of the spherical shell is concentrated at the center.

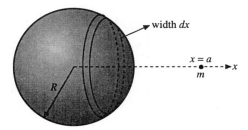

Hint: Divide the shell into rings of width dx that are perpendicular to the x-axis. Let x be the coordinate that represents the center of the typical ring (so x can range from $-R$ to R). In view of part 1, the mass of this ring is independent of x (it just depends on the width, dx). Use this to show that the mass of the ring is $M\,dx/2R$, and then use part 2 to show that the gravitational attraction between the mass m and this ring is

$$\frac{GmM}{2R}\left(\frac{a-x}{(a^2 + R^2 - 2ax)^{\frac{3}{2}}}\right)dx$$

Now, with the help of Maple, compute the gravitational attraction between m and the spherical shell, by integrating this expression from $x = -R$ to $x = R$. You will need to simplify your answer. Before doing this, type the command **assume(a>R,a>0,R>0);** which tells Maple that a and R are positive and that $a > R$.

4. In part 3, the mass m is assumed to be outside the shell. It is a fact that, if the mass is inside the shell, (i.e., if $a < R$), then the gravitational attraction between the mass m and the spherical shell is zero. Use Maple to verify this by re-executing the above integral but, this time, enter $a < R$ in your **assume** command.

5. Show that the gravitational attraction between a point mass m and a uniform ball of radius R and mass M is

$$\frac{GmM}{a^2}$$

where a is the distance between the mass m and the center of the ball. *This means that the gravitational attraction between the mass m and the ball is the same as if all the mass is concentrated at the center of the ball*—just as in the case of a spherical shell.

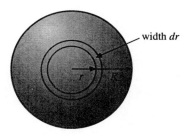

Hint: Let λ denote the (constant) mass density of the ball. Divide the ball into spherical shells of radius r with thickness dr. The mass of this shell is $4\pi\lambda r^2 dr$—i.e., (mass density) $\times$ (the volume of the shell). Use part 3 to find the gravitational attraction between the mass m and this spherical shell. Then integrate over r from $r = 0$ to $r = R$ and use the fact that $\lambda \times$ (the volume of the ball) $= M$. (This integral is simple enough that Maple is not required.)

6. For a very large mass such as the Earth, it is more realistic to treat the density λ as a function of r, the distance to the center of the ball. Show that the answer to the above problem is unchanged in this case. *Hint:* First express the mass M of the ball as an integral in terms of $\lambda = \lambda(r)$. You don't have to know a formula for $\lambda(r)$!

12.14 The Flight of a Baseball

Calculus Prerequisites: two-dimensional trajectory problems, velocity, acceleration, and differential equations. Read Sections 10.1, 10.3 and 10.4 in Stewart's **Calculus**.

Background: Imagine that a baseball player is up at home plate and hits the ball in the air. What parametric equations describe the position of the ball t seconds after it is hit? How far will the ball travel? How fast does the ball need to be hit in order for the ball to clear the home run fence? The following discussion and exercises are designed to answer these questions. Maple will help with the (extensive) computations and graphics.

First consider a simplified model that ignores air resistance. In this case, after the ball is hit, the only force acting on the ball is the vertical force due to gravity. Therefore, the following equations describe the x- and y- components of the acceleration of the ball

$$\frac{d^2x}{dt^2} = 0 \quad \text{and} \quad \frac{d^2y}{dt^2} = -g$$

where g is the acceleration constant due to gravity ($g = 32$, where the unit of distance is feet). Integrating these equations with respect to t gives

$$\frac{dx}{dt} = C_1 \quad \text{and} \quad \frac{dy}{dt} = -gt + C_2$$

where the constants C_1 and C_2 can be evaluated by considering the initial velocity of the ball. Suppose the initial speed of the ball is the constant v (in units of feet per second) and suppose the angle of inclination of the ball is A. Then the x- and y- components of the initial velocity of the ball are given by $v\cos(A)$ and $v\sin(A)$, respectively. After substituting $t = 0$ in the above equations and solving for C_1 and C_2, the following equations are obtained

$$\frac{dx}{dt} = v\cos(A) \quad \text{and} \quad \frac{dy}{dt} = -gt + v\sin(A)$$

One more integration with respect to t yields

$$x = vt\cos(A) + c_1 \quad \text{and} \quad y = \frac{-gt^2}{2} + vt\sin(A) + c_2$$

Assume that the origin ($x = 0, y = 0$) is the location of home plate and that the shoulder height of the batter (from where the ball is hit) is h feet. Then by substituting $t = 0$, the constants c_1 and c_2 can be found. The final parameterization of the baseball is given by

$$x = vt\cos(A) \quad \text{and} \quad y = \frac{-gt^2}{2} + vt\sin(A) + h$$

Assignment: Solve the following problems.

1. Set $g = 32$, the angle A at $\pi/4$, the velocity v to be 120 feet per second, and the height h to be 6 feet. Define the x- and y- coordinates of the ball t seconds later (see the equation for x and y above) as functions of t and then plot the trajectory of the ball until the ball hits level ground. Start the parameter t at 0, and experiment with different terminal values of t to try and get the entire flight of the baseball (until the ball hits the ground) on the screen. Find the horizontal distance traveled by the ball and the elapsed time.

2. What is the shape of the graph? Eliminate t, and determine the equation of the trajectory in the form of y as a function of x. To do this, solve for t in terms of x (this you can do by hand) and then use the **subs** command to substitute this expression for t into y.

   Now make the angle A and the initial speed v and t free variables **(A:='A'; v:='v'; t:='t';)** and re-input the equations for x and y into Maple.

3. Suppose the home run fence is 10 feet high and 350 feet from home plate. What is the minimum velocity at which the ball must leave the bat so that the ball barely clears the home run fence? *Be careful:* do not assume any particular value of the angle A. In fact your strategy should be as follows. First, eliminate t as you did in part 2 (only now using v and A as free variables). Then solve for the speed v in terms of the angle A by substituting $x = 350$, $y = 10$ and solving the resulting equation. Finally, minimize v as a function of A.

4. Now assume air resistance acts on the ball. Air resistance acts in the opposite direction to the velocity of the ball and its magnitude is proportional to the speed. This leads to the acceleration equations

$$\frac{d^2x}{dt^2} = -k\frac{dx}{dt} \quad \text{and} \quad \frac{d^2y}{dt^2} = -g - k\frac{dy}{dt}.$$

   Here, k is a friction constant, which will be given later. Solve these equations for x and y (using the initial conditions). Take the limit of your solutions as $k \rightarrow 0$ and see if your result agrees with the solution for x and y without air resistance. Then repeat parts 1 and 3, taking into account air resistance with $k = 0.1$. Compare your plots and your answers to your results without air resistance. For part 3, you will not be able to algebraically solve for v in terms of A (even with Maple). Instead, take specific values of A (near $\pi/4$) and solve for v numerically (using **fsolve**). Determine an approximate minimum value for v.

Related Activities: See the Applied Projects: *Which is Faster, Going Up of Coming Down?* (at the end of Section 10.3) and *Calculus and Baseball* (at the end of Section 10.4) in Stewart's text.

12.15 Logistic Growth

Calculus Prerequisite: first order separable differential equations. Read Section 10.3 in Stewart's **Calculus.**

Background: The logistic equation is a first order nonlinear differential equation of the form

$$\frac{dP}{dt} = aP - bP^2$$

which is often used to study population growth. The parameters a and b are positive constants that are related to the birth and death rate of the population. In the problem that follows, $P(t)$ represents the population of the United States at time t, in years.

The table below contains population data for the United States with population values given in millions. **Re-scale the time variable so that $t = 0$ corresponds to the year 1790.**

Year	Population	Year	Population
1790	3.93	1900	75.99
1800	5.3331	1910	91.97
1810	7.24	1920	105.71
1820	9.64	1930	122.78
1830	12.87	1940	131.67
1840	17.07	1950	151.333
1850	23.19	1960	179.32
1860	31.44	1970	203.21
1870	39.82	1980	226.5
1880	50.16	1990	249.63
1890	62.95	2000	?

Assignment: The goal of this project is to find positive constants a and b such that the solution to the logistic differential equation above satisfying $P(0) = 3.93$ *best fits* the data in the table above. One method for achieving a *best fit* is to proceed as follows. Begin by naming the population values in the table above P_0, P_1, ... , P_{20}. That is, denote $P_0 = 3.93$, $P_1 = 5.3331$, etc. Now, find the solution $P(t)$ of the logistic equation above satisfying $P(0) = 3.93$ as a function of a, b and t, and consider the function given by

$$F(a,b) = \sum_{n=1}^{20} (P(10n) - P_n)^2 .$$

Determine a mechanism for finding the positive values of a and b which minimize $F(a,b)$.

Use these values of a and b to estimate the population of the United States in the year 2000. Do you think this is a reasonable model for predicting the population in the year 2000?

12.16   Radioactive Waste at a Nuclear Power Plant

Calculus Prerequisites: first order, linear differential equations. Read Section 10.6 in Stewart's **Calculus**.

Background: A nuclear power plant produces a waste product that is a radioactive isotope, called A. The isotope A has a half-life of 10 years; 70% by weight decays into a radioactive isotope B, and 30% into a radioactive isotope C. The isotope B has a half-life of 20 years and decays into nonradioactive by-products. The radioactive isotope C has a half-life of 30 years and also decays into nonradioactive by-products.

Assume that isotopes B and C weigh essentially the same as isotope A. Thus, for example, if 100 kilograms of A decays there will be 70 kilograms of B and 30 kilograms of C.

Assignment: Answer the following questions.

1. Suppose you start with 400 kilograms of isotope A. What is the maximum amount of isotope B that will be present, and when will this occur? Answer the same question for isotope C.

2. When the power plant was first turned on, there was no isotope A, B, or C present. If the power plant operates so that it produces isotope A at the constant rate of 40 kilograms per year, what are the maximum amounts of isotopes A, B, and C that will be present, and when will these different maxima occur?

3. Federal safety requirements say that the reactor can never have on hand more than 500 kilograms of isotope A, 400 kilograms of isotope B, or 300 kilograms of isotope C. What is the maximum rate at which the power plant can produce isotope A without violating the federal regulations?

Hints:

(a) First find the three decay constants.

(b) If a radioactive isotope is being produced by some source at the same time as it is decaying, how does that alter the differential equation for the rate of change of the amount of this isotope?

(c) Be sure to plot A, B, and C, in order to ascertain if they have the qualitative behavior you expect.

12.17   Pension Funds

Calculus Prerequisites: linear differential equations and exponential growth. Read Sections 10.4 and 10.6 in Stewart's **Calculus**.

Background: A pension fund starts out with P (at $t = 0$) and is invested with a return of $100r\%$ per year, compounded continuously (here, r is the interest rate, given as a number between 0 and 1). The pension fund must continuously pay out money at the rate of R per year to its employees for a period of n years (this means that the value of the pension fund decreases to zero after n years). Let $y(t)$ denote the value of the pension fund after t years.

Assignment: Answer the following questions with complete sentences.

1. From the information given, derive the differential equation $y' = ry - R$, with the conditions $y(0) = P$, $y(n) = 0$.

2. Solve this differential equation for y.

3. Find a formula for P in terms of r, n, and R (P represents the amount of money required to pay out R per year for n years, assuming the rate of return on the investment is $100r\%$).

4. Calculate P for $R = 50K$, $r = .07$, and $n = 20$.

5. Calculate the interest rate (r) required so that an initial value of $P = 500K$ for the pension fund will pay out $50K$ per year for 30 years.

Hints: Maple is not needed in an essential way for parts 1-4. However, you can check your solution to the differential equation (part 2) by using Maple's **dsolve** command for solving differential equations. To use this command, first enter the differential equation and give it a label (such as *eq*).

```
> eq:=diff(y(t),t)=r*y(t)-R;
```

Then, to solve this differential equation, issue the command

```
> dsolve(eq,y(t));
```

The general solution to the differential equation is returned, with an unknown constant of integration. This constant can be evaluated using the initial condition $y(0) = P$. The differential equation with the initial condition can be solved with the command

```
> dsolve({eq,y(0)=P},y(t));
```

In part 5, Maple is necessary to solve the relevant equation for r.

12.18   Visualizing Euler's Method

Calculus Prerequisites: minimal exposure to numerical methods for differential equations. Read Section 10.2 in Stewart's **Calculus**.

Background: The goal of this lab is to use Maple to illustrate Euler's method, and to show how Euler's method leads naturally to numerical anti-differentiation, numerical integration, and linear splines.

Euler's method for approximating the solution to the problem

$$\frac{dy}{dt} = f(t, y), \quad y(t_0) = y_0 \tag{12.1}$$

is as follows: choose a step size $h > 0$, and let $t_i = t_0 + ih$, $i = 0, 1, 2, ..., n$, where n is the number of steps of size h you must take to reach some prescribed final value of t. Then, recursively compute the values

$$y_{i+1} = y_i + h\, f(t_i, y_i), \quad i = 0, 1, 2, ..., n - 1. \tag{12.2}$$

The y_i's are intended to be approximations to the true solution, $y(t)$, at the discrete points t_i; that is, $y_i \approx y(t_i)$ for $i = 0, 1, ..., n$. To see where the method comes from, observe that the derivative of the true solution at t_0 is $y'(t_0) = f(t_0, y(t_0)) = f(t_0, y_0)$, as follows from Eq. (12.1). Hence, the line tangent to the graph of $y(t)$ at the point (t_0, y_0) has the equation $y = y_0 + (t - t_0)f(t_0, y_0)$. If we use this as an approximation to the true solution for t near t_0, then when we let $t = t_0 + h = t_1$, we obtain the approximation $y_1 = y_0 + h\, f(t_0, y_0)$ for $y(t_1)$. This is Eq. (12.2) with $i = 0$. The algorithm in Eq. (2) is just a continuation of this procedure.

Below we show one way of implementing Euler's method with Maple, using a procedure called **Euler**. We first load two packages we will need for plotting.

```
> with(plots):
> with(DEtools):
```

The procedure **Euler** is as follows. The output is a list containing the values $y_0, y_1, ..., y_n$.

```
> Euler := proc(f,t0,y0,h,n)
> local i,y;
> y[0] := y0;
> for i from 0 to n-1 do
> y[i+1] := evalf(y[i]+h*f(t0+i*h,y[i]));
> od;
> y;
> end;
```

The following example will illustrate how this procedure is used in a typical problem, and how to plot the results. In Eq. (12.1), we let $f(t, y) = y$, $t_0 = 0$, and $y_0 = 1$. The exact solution in this case is easily seen to be $y(t) = e^t$. We will also plot the direction field and the true solution, and display all three plots together.

```
> f :=  (t,y) -> y:
> t0 := 0:
> y0 := 1:
> n := 4:
> h := evalf(1/n):
> e1 := Euler(f,t0,y0,h,n):
> e1list := NULL:
> for i from 0 to n do
> ti := evalf(t0+i*h):
> print(ti,e1[i]);
> e1list := e1list,ti,e1[i]:
> od:
> p1 := plot( [e1list],0..1,0..3,style=line):
> p2 := plot( [e1list],0..1,0..3,style=point):
> display({p1,p2});
```

We next include the direction field.

```
> p3 := dfieldplot(f(t,y),[t,y],0..1,0..3,arrows = LINE,grid=[20,20]):
> display({p1,p2,p3});
```

The purpose of this plot is to show how the straight line segments in the Euler solution have slopes that agree with the direction field at the *left* endpoint of the line segment. You will be asked in the exercises to explain this. Finally, we include the true solution, to get an idea of how good the approximation is.

```
> p4 := plot(exp(t),t=0..1,0..3):
> display({p1,p2,p3,p4});
```

Assignment:

1. Apply Euler's method with $n = 2$ to the problem

$$\frac{dy}{dt} = f(t,y) = 2\sin\left(1 + 4\,e^t\right) \equiv g(t), \quad y(0) = 0, \quad 0 \le t \le 1 \qquad (12.3)$$

2. What is the exact solution to Eq. (12.3)? *Hint:* It's okay to give your answer as an integral, with the variable upper limit t. How is this solution related to the set of antiderivatives of $g(t)$?

3. Use your Euler solution to obtain an approximation to the definite integral

$$\int_0^1 g(t)\,dt$$

Compare this with the answer obtained using Maple's **int** command.

4. Plot the direction field and the Euler solution together, in the plot window $0 \le t \le 1, \ -1 \le y \le 1$.

5. Find the piecewise linear function $s(t)$ that goes through the points $\{(t_i, y_i)\}_{i=0}^{n}$. For example, the piecewise linear function that goes through the points $(0,0)$, $(0.5,1)$, $(1,0)$ is

$$s(t) = \begin{cases} 2t & \text{if } 0 \le t \le 0.5 \\ 2 - 2t & \text{if } 0.5 < t \le 1 \end{cases}$$

Next, use on-line help to find out how to use Maple's **spline** command to obtain $s(t)$. Compare your answer to Maple's.

6. How are the slopes of the straight line segments in the graph of the Euler solution related to the direction field? Explain the reason for this relation.

7. Repeat Exercises 1, 3, and 4 with $n = 4$.

8. Euler's method is usually too inaccurate to be of practical use, but it is a good starting point for learning about numerical methods for differential equations. A significant improvement is the following, called the modified Euler's method (or Huen's method, or second order Runge-Kutta method)

$$y_{i+1} = y_i + \frac{h}{2}(k_1 + k_2), \quad \text{where } \ k_1 = f(t_i, y_i) \ \text{ and } \ k_2 = f(t_i + h, y_i + hk_1)$$

Write a procedure called **Eulermod** to implement this method, and apply it to the example (i.e., with solution $y = e^t$), using $n = 2$. Display plots showing the approximate solution, together with the exact solution. Repeat for $n = 4$.

12.19   The Brightest Phase of Venus

Calculus Prerequisites: This is the hardest project - nontrivial trigonometry, max/min theory, area. Read Sections 4.7, 5.1 and 6.1 in Stewart's **Calculus**.

Background: The brightness of Venus is proportional to the area of the visible portion of Venus and inversely proportional to the square of the distance from the Earth to Venus. From the accompanying figure, note that, as the angle t increases from 0 to π, the area of the visible portion of Venus increases. This tends to increase the brightness of Venus. But the distance d from the Earth to Venus also increases, which tends to decrease the brightness of Venus. For some angle t, between 0 and π, Venus will appear brightest.

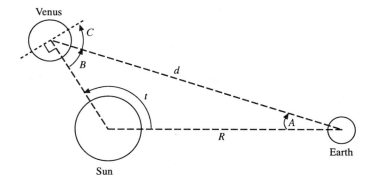

Assignment: Find the brightest phase of Venus (i.e., the angle t) by following the outline given below. Write up your solution using complete sentences. Use Maple where appropriate to help with the computations and graphics.

1. As mentioned above, brightness is proportional to the quantity

$$\frac{\text{Area of visible portion of Venus}}{d^2}$$

2. From geometry, show that $B = \pi - (t + A)$ and $C = (t + A) - \pi/2$.

3. Let a be the radius of Venus. Show that the visible portion of Venus (from Earth) lies between the curves $x = -\sqrt{a^2 - y^2}$ and $x = \sin(C)\sqrt{a^2 - y^2}$. Now show that the area of the visible portion of Venus is

$$\frac{\pi a^2}{2}(1 + \sin(C))$$

4. Combine parts 1, 2, and 3 to obtain a formula for the brightness of Venus that depends on the angles t and A and the distance d. The goal is to express the brightness in terms of one variable t. To this end, use the law of cosines and the law of sines to show the following

$$d^2 = r^2 + R^2 - 2rR\cos(t)$$

$$\sin(A) = \frac{r\sin(t)}{d}$$

Here, r is the distance from the Sun to Venus and R is the distance from the Sun to the Earth. Use the values $R = 93$, $r = 67$, and $a = 0.004$ (the unit of distance is a million miles).

5. Use the equations in part 4 to find an expression for the brightness of Venus that depends on one variable t. Use Maple to maximize this function over the interval $0 \le t \le \pi$.

Index